Mi
the Nutritional Impacts of the Global Food Price Crisis

Workshop Summary

Elizabeth Haytmanek and Katherine McClure, *Rapporteurs*

Board on Global Health

Food and Nutrition Board

INSTITUTE OF MEDICINE
OF THE NATIONAL ACADEMIES

THE NATIONAL ACADEMIES PRESS
Washington, D.C.
www.nap.edu

THE NATIONAL ACADEMIES PRESS 500 Fifth Street, N.W. Washington, DC 20001

NOTICE: The project that is the subject of this report was approved by the Governing Board of the National Research Council, whose members are drawn from the councils of the National Academy of Sciences, the National Academy of Engineering, and the Institute of Medicine.

This study was supported by Grant No. 51649 between the National Academy of Sciences and the Bill & Melinda Gates Foundation, with additional support from the PepsiCo Foundation. Any opinions, findings, conclusions, or recommendations expressed in this publication are those of the author(s) and do not necessarily reflect the view of the organizations or agencies that provided support for this project.

International Standard Book Number-13:   978-0-309-14018-8
International Standard Book Number-10:   0-309-14018-8

Additional copies of this report are available from the National Academies Press, 500 Fifth Street, N.W., Lockbox 285, Washington, DC, 20055; (800) 624-6242 or (202) 334-3313 (in the Washington metropolitan area); Internet, http://www.nap.edu.

For more information about the Institute of Medicine, visit the IOM home page at: **www.iom.edu.**

Copyright 2010 by the National Academy of Sciences. All rights reserved.

Printed in the United States of America

The serpent has been a symbol of long life, healing, and knowledge among almost all cultures and religions since the beginning of recorded history. The serpent adopted as a logotype by the Institute of Medicine is a relief carving from ancient Greece, now held by the Staatliche Museen in Berlin.

Suggested citation: IOM (Institute of Medicine). 2010. *Mitigating the Nutritional Impacts of the Global Food Price Crisis: Workshop Summary.* Washington, DC: The National Academies Press.

*"Knowing is not enough; we must apply.
Willing is not enough; we must do."*
—Goethe

INSTITUTE OF MEDICINE
OF THE NATIONAL ACADEMIES

Advising the Nation. Improving Health.

THE NATIONAL ACADEMIES
Advisers to the Nation on Science, Engineering, and Medicine

The **National Academy of Sciences** is a private, nonprofit, self-perpetuating society of distinguished scholars engaged in scientific and engineering research, dedicated to the furtherance of science and technology and to their use for the general welfare. Upon the authority of the charter granted to it by the Congress in 1863, the Academy has a mandate that requires it to advise the federal government on scientific and technical matters. Dr. Ralph J. Cicerone is president of the National Academy of Sciences.

The **National Academy of Engineering** was established in 1964, under the charter of the National Academy of Sciences, as a parallel organization of outstanding engineers. It is autonomous in its administration and in the selection of its members, sharing with the National Academy of Sciences the responsibility for advising the federal government. The National Academy of Engineering also sponsors engineering programs aimed at meeting national needs, encourages education and research, and recognizes the superior achievements of engineers. Dr. Charles M. Vest is president of the National Academy of Engineering.

The **Institute of Medicine** was established in 1970 by the National Academy of Sciences to secure the services of eminent members of appropriate professions in the examination of policy matters pertaining to the health of the public. The Institute acts under the responsibility given to the National Academy of Sciences by its congressional charter to be an adviser to the federal government and, upon its own initiative, to identify issues of medical care, research, and education. Dr. Harvey V. Fineberg is president of the Institute of Medicine.

The **National Research Council** was organized by the National Academy of Sciences in 1916 to associate the broad community of science and technology with the Academy's purposes of furthering knowledge and advising the federal government. Functioning in accordance with general policies determined by the Academy, the Council has become the principal operating agency of both the National Academy of Sciences and the National Academy of Engineering in providing services to the government, the public, and the scientific and engineering communities. The Council is administered jointly by both Academies and the Institute of Medicine. Dr. Ralph J. Cicerone and Dr. Charles M. Vest are chair and vice chair, respectively, of the National Research Council.

www.national-academies.org

PLANNING COMMITTEE ON MITIGATING THE NUTRITIONAL IMPACTS OF THE GLOBAL FOOD PRICE CRISIS[1]

REYNALDO MARTORELL (*Chair*), Robert W. Woodruff Professor, International Nutrition; Senior Advisor, Global Health Institute, Hubert Department of Global Health, The Rollins School of Public Health, Emory University, Atlanta, Georgia
HANS HERREN, President, Millennium Institute, Arlington, Virginia
ISATOU JALLOW, Chief, Women, Children and Gender Policy, UN World Food Program, Rome, Italy
RUTH K. ONIANG'O, Executive Director, Rural Outreach Program, Nairobi, Kenya
PER PINSTRUP-ANDERSEN, H.E. Babcock Professor of Food, Nutrition and Public Policy, Cornell University, Ithaca, New York
JUAN A. RIVERA, Director, Center for Research in Nutrition and Health, National Institute of Public Health, Mexico; Professor, Nutrition, School of Public Health, Cuernavaca, Mexico
RICARDO UAUY, Professor, Nutrition and Pediatrics, University of Chile, Santiago, Chile
KEITH P. WEST, JR., Professor, International Nutrition, Bloomberg School of Public Health, Johns Hopkins University, Baltimore, Maryland

Study Staff

ELIZABETH HAYTMANEK, Study Director
KATHERINE McCLURE, Senior Program Associate
GUI LIU, Senior Program Assistant
MEGAN PEREZ, Intern
JULIE WILTSHIRE, Financial Officer
PATRICK KELLEY, Director, Board on Global Health
LINDA D. MEYERS, Director, Food and Nutrition Board

---

[1] IOM planning committees are solely responsible for organizing the workshop, identifying topics, and choosing speakers. The responsibility for the published workshop summary rests with the workshop rapporteurs and the institution. The planning committee's role was limited to planning the workshop, and the workshop summary has been prepared by the workshop rapporteurs as a factual summary of what occurred at the workshop.

Reviewers

This report has been reviewed in draft form by individuals chosen for their diverse perspectives and technical expertise, in accordance with procedures approved by the National Research Council's Report Review Committee. The purpose of this independent review is to provide candid and critical comments that will assist the institution in making its published report as sound as possible and to ensure that the report meets institutional standards for objectivity, evidence, and responsiveness to the study charge. The review comments and draft manuscript remain confidential to protect the integrity of the process. We wish to thank the following individuals for their review of this report:

Eileen Kennedy, Tufts University Friedman School of Nutrition Science and Policy
Vivica Kraak, Save the Children
Per Pinstrup-Andersen, Cornell University
Meera Shekar, The World Bank

Although the reviewers listed above have provided many constructive comments and suggestions, they were not asked to endorse the final draft of the report before its release. The review of this report was overseen by **Hugh Tilson,** University of North Carolina, Gillings School of Global Public Health. Appointed by the Institute of Medicine, he was responsible for making certain that an independent examination of this report was carried out in accordance with institutional procedures and that all review comments were carefully considered. Responsibility for the final content of this report rests entirely with the authors and the institution.

Contents

SUMMARY ... 1

1 INTRODUCTION ... 7
Workshop Background, 8
Welcome from the Sponsor, 9
   Ellen Piwoz
References, 12

2 THE DUAL CRISES: TANDEM THREATS TO NUTRITION ... 13
The Recent and Current Food Price Crisis and Future Perspectives, 14
   Per Pinstrup-Andersen
The Current Global Economic Crisis and Future Perspectives, 21
   Hans Timmer
Discussion, 26
References, 30

3 IMPACTS ON NUTRITION ... 31
Conceptual Presentation on Pathways to Nutritional Impact, 31
   Ricardo Uauy
Existing Evidence of Nutritional Impacts, 38
   Francesco Branca
Are the Urban Poor Particularly Vulnerable?, 43
   Marie Ruel
Discussion, 46
References, 48

4   RESPONDING TO THE CRISES AT THE COUNTRY LEVEL   49
    The Role of Ministries in Responding to the Crises at the
    Country Level, 49
        Ruth Oniang'o
    Review of National Responses to the Food Crisis, 51
        Hafez Ghanem
    The Case of Mexico, 55
        Graciela Teruel Belismelis
    The Global Food Price Crisis and Food Development Strategy in China, 57
        Fangquan Mei
    Food Prices, Consumption, and Nutrition in Ethiopia: Implications of
    Recent Price Shocks, 60
        Paul Dorosh
    Bangladesh Case Study, 66
        Josephine Iziku Ippe
    Discussion, 71
    References, 72

5   A ROLE FOR NUTRITION SURVEILLANCE IN ADDRESSING THE
    GLOBAL FOOD CRISIS   75
    Nutrition Surveillance in Relation to the Food Price and
    Economic Crises, 76
        John Mason
    Insights from 25 Years of Helen Keller International's Nutrition
    Surveillance in Bangladesh and Indonesia, 82
        Andrew Thorne-Lyman
    Famine Early Warning Systems Network, Nutrition Surveillance, and
    Early Warning, 85
        Chris Hillbruner
    Listening Posts Project: A Concept for a Real-Time Surveillance
    System Nested Within a Program, 88
        Anna Taylor
    Food Security, Nutrition Monitoring, and the Global Food Price Crisis:
    USAID/FFP Title II Programs, 91
        Ellen Mathys
    Discussion, 95
    References, 96

6   THE GLOBAL RESPONSE TO THE CRISES   99
    Introduction to the Global Nutrition Landscape, 99
        Ruth Levine
    The Role and Capacity of Foundations in Responding to the Crises, 106
        Haddis Tadesse

The Role of Food Companies in Responding to the Crises, 110
    Derek Yach
The Advocacy Role of Civil Society Organizations in Responding to the Economic and Food Price Crises, 114
    Asma Lateef
The Role and Capacity of Civil Society in Responding to the Crises, 116
    Tom Arnold
Mitigating the Nutritional Impact of the Global Food Security Crisis: The Role and Capacity of UN Agencies in Response to the Crisis, 118
    David Nabarro
The Role and Capacity of UNICEF in Responding to the Crises, 123
    Werner Schultink
The Role and Capacity of the WFP in Responding to the Crises, 125
    Martin Bloem
The Role and Capacity of FAO in Responding to the Crises, 127
    Hafez Ghanem
The Role and Capacity of WHO in Responding to the Crises, 129
    Francesco Branca
Discussion, 131
References, 133

7 U.S. POLICY IN FOOD AND NUTRITION   135
    The Roadmap to End Global Hunger, 135
        James McGovern
    USAID's Response to the Food Crisis and Preventing Malnutrition for the Future, 140
        Michael Zeilinger
    Food Security in the 21st Century, 145
        Nina Fedoroff
    USDA's Response to the Crises and Future Perspectives, 147
        Rajiv Shah
    Renewing American Leadership in the Fight Against Global Hunger and Poverty, 151
        Catherine Bertini and Dan Glickman
    Discussion, 155
    Workshop Closing Remarks, 156
        Reynaldo Martorell
    References, 158

APPENDIXES
A Workshop Agenda   159
B Speaker Biographies   165
C Workshop Registrants   183

Summary

In 2007 and 2008, the world witnessed a dramatic increase in food prices. High food prices caused civil unrest and exacerbated the humanitarian crisis of food insecurity and malnutrition. The global financial crisis that began in 2008 compounded the burden of high food prices, exacerbating the problems of hunger and malnutrition in developing countries. The tandem food price and economic crises struck amidst the massive, *chronic* problem of hunger and undernutrition in developing countries. The nutritional consequences of the food price increases, compounded by the economic downturn, could be considerable in poor urban populations, rural areas that are net food purchasers, and in female-headed households. Malnutrition affects the survival, health, well-being, and developmental potential of vulnerable groups.

National governments and international actors have taken a variety of steps to mitigate the negative effects of increased food prices on particular groups. Emphasizing the importance of child and maternal health and nutrition to international development, the Millennium Development Goals (MDGs) represent a global commitment to poverty and hunger eradication.[1] The recent abrupt increase in food prices, in tandem with the current global economic crisis, threatens progress made in these areas and could prove a serious barrier to achieving the MDGs.

---

[1] The MDGs were adopted in 2000 by the member nations of the United Nations and the world's major development institutions.

WORKSHOP BACKGROUND

The Institute of Medicine (IOM), with funding from the Bill & Melinda Gates Foundation and the PepsiCo Foundation, held a workshop titled *Mitigating the Nutritional Impacts of the Global Food Price Crisis* on July 14–16, 2009, in Washington, DC, at the Kaiser Family Foundation's Barbara Jordan Conference Center. Presenters were chosen by a planning committee to describe the dynamic technological, agricultural, and economic issues contributing to the food price increases of 2007 and 2008 and their impacts on health and nutrition in resource-poor regions. The planning committee quickly realized that it was impossible to ignore the compounding effects of the current global economic downturn on nutrition. Subject matter experts were invited to the workshop and asked to discuss these tandem crises, their impacts on the nutritional status of vulnerable populations, and opportunities to mitigate their negative nutritional effects.

The planning committee's role was limited to planning the workshop, and the workshop summary has been prepared by the workshop rapporteurs as a factual summary of what occurred at the workshop. The reader should be aware that the material presented here expresses the views and opinions of the individuals participating in the workshop and not the deliberations and conclusions of a formally constituted IOM consensus study committee. These proceedings summarize only what participants stated in the workshop and are not intended to be an exhaustive exploration of the subject matter and should not be perceived as a consensus of the participants, nor the views of the planning committee, the IOM, or its sponsors.

THE DUAL CRISES: TANDEM THREATS TO NUTRITION

A strong evidence base underpinning and motivating investment in nutrition of vulnerable populations has emerged over the past 5 years. Specifically, a *Lancet* series, published in January 2008, clearly showed that maternal and child undernutrition is the underlying cause of 3.5 million deaths, 35 percent of the disease burden in children younger than 5 years, and 11 percent of total global disability-adjusted life years (DALYs) (Black et al., 2008). With this foundation, it seemed that the will, the tools, and the technologies had all been mobilized, and real progress in nutrition could be made. Then the sudden increases in global food prices in 2007 and 2008, exacerbated by the current global economic downturn, began to threaten the hoped-for trajectory of progress. Between March 2007 and March 2008, price rises of 31 percent for corn, 74 percent for rice, 87 percent for soya, and 130 percent for wheat were documented (Hawtin, 2008).

Per Pinstrup-Andersen explained that price *volatility* and rapid food price *fluctuations* most significantly affect the global poor. While higher food prices are of course problematic, if they were consistent and predictable, food buyers could better cope. Dr. Pinstrup-Andersen forewarned that such price volatility is predicted to increase in the future. The World Bank's Hans Timmer forecasted that more than half of the global economic recovery from the current downturn will come from developing countries. In this sense, he argued, protecting and promoting the growth of developing countries' economies serves the interests of rich and poor countries alike.

IMPACTS ON NUTRITION

The food price and economic crises will have both short-term and long-term impacts on the nutritional status of vulnerable populations. The global poor are usually the hardest hit by food price increases and economic strife. At the household level in developing countries, poor consumers spend 50 to 70 percent of their budget on food (von Braun et al., 2008), so their capacity to absorb rises in prices (or lowered incomes) is limited and often forces difficult household choices that adversely affect women and children in particular.

Ricardo Uauy noted that decreasing household income has a disproportionate effect on micronutrient intake and thus the *quality* of diets, rather than quantity; families forced to feed themselves with less purchasing power tend not to decrease the staples in their diets (rice or grain), but instead to eliminate the vegetables and animal products—which contain essential micronutrients. Dr. Uauy also predicted that economic growth will be restricted by the reduced productivity of children who are suffering the negative impacts of the global food crisis, which will have a long-term, transformational effect upon society's development. Marie Ruel explained that poverty itself is a strong indicator of how people will be affected by soaring food prices. A large proportion of poor people—in both urban and rural areas—are "net buyers" of food and therefore need money to purchase food. Dr. Ruel further noted that female-headed households are the most vulnerable of all and may be forced to cope in ways that deinvest in children (e.g., taking children out of school) and have lifelong effects on those children's development and future earning potential.

RESPONDING TO THE CRISES AT THE COUNTRY LEVEL

Between 2003 and 2008, the world prices of maize and wheat tripled and the price of rice quadrupled (von Braun, 2008). Individual countries dealt with these dramatic food price spikes in a range of ways. In evaluating how the food crisis affected food security and nutrition to varying degrees in different countries and regions, Hafez Ghanem of the Food and Agriculture Organization argued that the term *crisis* is a misnomer. The billion hungry people in the world signify a chronic, structural problem in the global food and agriculture system. According to Mr. Ghanem, because there has been no public outcry, the problem of chronic hunger receives no political attention. Ruth Oniang'o lamented that country governments reacted to the riots and demonstrations during the food price spikes in order to protect themselves and maintain public safety, not to tackle the broader, systemic issues of poverty and malnutrition.

A ROLE FOR NUTRITION SURVEILLANCE IN ADDRESSING THE GLOBAL FOOD CRISIS

The food price and economic crises have highlighted the need for data collection in order to understand the effects of these phenomena on populations. There are a variety of tactical measures and approaches to nutrition surveillance at work in different countries and regions. A number of presenters spoke of the need to aggregate data, compile it quickly using new technologies, and deliver it to the food security and nutrition community for decision making at the program and policy level. Several workshop participants emphasized the need to develop in-country capacity (of governments and nongovernmental organizations) to conduct their own surveillance in order to ensure local acceptance and use of the data collected.

Nutrition surveillance mechanisms can play a role in predicting and preventing future crises, as well as documenting the impacts of crises to inform programs and policies that work to ameliorate the negative nutritional impacts of food crises. The effects of undernutrition are known, and the 2008 *Lancet* series describes effective strategies for mitigation (Bhutta et al., 2008). The hungry and malnourished deserve such evidence-based action to alleviate their circumstances.

THE GLOBAL RESPONSE TO THE CRISES

A broad group of people and organizations work in the nutrition landscape. These include multilateral UN agencies, bilateral government agencies, nongovernmental organizations, universities, research institutions, foundations, and the private sector. The people and organizations who work in the field of nutrition have varying roles, functions, and capacities to deal with the outcomes of the recent food price and current economic crises.

There are a number of new players in food security and nutrition who require leadership and coordinating mechanisms to function efficiently and without overlap. Ruth Levine recommended that a high-level mandate for institutional change should set the expectations for how institutions should allocate roles and work together. Dr. Levine also stated that additional resources for bolstering institutional capacity within the UN and other agencies would be needed to respond to such a mandate and that serious engagement of the private sector should be fostered. Dr. Levine and other speakers urged the spectrum of players in the nutrition landscape to heed the call for major structural changes on a sustainable, long-term basis, because short-term, emergency actions are not sufficient in dealing with the growing numbers of global hungry.

U.S. POLICY IN FOOD AND NUTRITION

The U.S. government can play an important role in the fight to end global hunger, and there is a renewed sense of political will to address these issues. U.S. Congressman James McGovern presented the *Roadmap to End Global Hunger*—a coalition of more than 30 nonprofit organizations—that calls for a comprehensive government-wide approach to alleviate global hunger and promote food and nutrition security (*Roadmap to End Global Hunger*, 2009). Rajiv Shah described how the U.S. Department of Agriculture is expanding its resources in technology development in promising areas like food biofortification, pest and disease resistant vegetable breeding, and livestock improvement, with hopes of extending these technologies to smallholder farmers in developing countries. Catherine Bertini and Dan Glickman presented the Chicago Council on Global Affairs' report, *Renewing American Leadership in the Fight Against Global Hunger and Poverty,* which aims to put agricultural development back at the center of U.S. development policy.

THE WAY FORWARD—THEMES FROM THE WORKSHOP

The following themes emerged during the workshop through several speakers' presentations and during discussion sessions with workshop participants. These themes are not intended to be and should not be perceived as a consensus of the participants, nor the views of the planning committee, the IOM, or its sponsors.

- The current crisis presents an opportunity to motivate donors and engage affected country governments in efforts to address undernutrition, hunger, and food insecurity in vulnerable populations.
- There is a window of opportunity with women and children where known nutritional interventions will be most effective and have a long-term payoff, as described in the 2008 *Lancet* series on maternal and child undernutrition.
- There is a simultaneous call for better quality data to inform program design and effectiveness, but there is also a critical need to immediately move forward with proven programs and policies to mitigate hunger and undernutrition in vulnerable populations.
- Short-term, emergency actions are not sufficient to remedy recurring food crises; instead, both short- and long-term investments in global food and agriculture systems are needed.
- Mechanisms to help vulnerable populations cope with food price volatility and to prevent future shocks are required.
- It is important to draw upon the expertise of governments, nongovernmental organizations (NGOs) and civil society, the private sector, foundations, and the broad spectrum of actors in the international nutrition and agriculture sectors.

- The roles of the multiple UN agencies that work to promote the food and nutrition security of vulnerable populations need to be clarified.
- Fostering engagement with the private sector may yield new expertise and resources.
- A stronger voice from indigenous non-governmental organizations is needed. Such local non-governmental organizations could benefit from capacity-building efforts to encourage ownership and political involvement.

REFERENCES

Bhutta, Z. A., T. Ahmed, R. E. Black, S. Cousens, K. Dewey, E. Giugliani, B. A. Haider, B. Kirkwood, S. S. Morris, H. Sachdev, and M. Shekar. 2008. What works? Interventions for maternal and child undernutrition and survival. *Lancet* 371(9610):417-440.

Black, R. E., L. H. Allen, Z. A. Bhutta, L. E. Caulfield, M. de Onis, M. Ezzati, C. Mathers, and J. Rivera. 2008. Maternal and child undernutrition: Global and regional exposures and health consequences. *Lancet* 371(9608):243-260.

Hawtin, G. 2008. *CIAT's Response to the World Food Situation*. Cali, Columbia: CIAT.

Roadmap to End Global Hunger. 2009. Washington, DC.

von Braun, J. 2008. *Food and Financial Crises: Implications for Agriculture and the Poor.* Washington, DC: International Food Policy Research Institute.

von Braun, J., A. Ahmed, K. Asenso-Okyere, S. Fan, A. Gulati, J. Hoddinott, R. Pandya-Lorch, M. W. Rosegrant, M. Ruel, M. Torero, T. van Rheenen, and K. von Grebmer. 2008. *High Food Prices: The What, Who, and How of Proposed Policy Actions.* Washington, DC: International Food Policy Research Institute.

1

Introduction

In 2007 and 2008, the world witnessed a dramatic increase in food prices. Between March 2007 and March 2008, price increases of 31 percent for corn, 74 percent for rice, 87 percent for soya, and 130 percent for wheat were documented (Hawtin, 2008). This increase in food prices posed a heavy burden on consumers in food-importing countries. The pressure of increasing food prices was a major factor in riots that erupted in many countries (Ngongi, 2008). High food prices not only caused civil unrest, but also exacerbated the humanitarian crisis of food insecurity; the tandem food price and economic crises struck amidst the massive, *chronic* problem of hunger and undernutrition in developing countries. Soaring food and fuel prices are a "perfect storm" for the most vulnerable billion—those living on $1 a day—who can't afford to see their purchasing power further decrease (Sheeran, 2008).

The nutritional consequences of the food price increases could be considerable in poor urban populations, in rural areas that are net food purchasers, and in female-headed households. Malnutrition affects the survival, health, well-being, and developmental potential of vulnerable groups. Food shortages disproportionately impact women during pregnancy, leading to irreversible physical and cognitive damage to their unborn child. Both quality and quantity of the diet are important for successful birth outcomes (*Global recession increases malnutrition for the most vulnerable people in developing countries*, 2009).

National governments and international actors have taken a variety of steps to mitigate the effects of increased food prices on particular groups. While some of these actions have helped stabilize food prices, other actions may help certain groups at the expense of others, make food prices more volatile in the long run, and distort trade markets (von Braun, 2008).

Emphasizing the importance of child and maternal health and nutrition to international development, the Millennium Development Goals (MDGs) represent a global commitment to poverty and hunger eradication.[1] The first MDG is to eradicate extreme hunger and poverty; one of the 3 targets of MDG 1 is to halve, between 1990 and 2015, the proportion of people who suffer from hunger. MDGs 4 and 5 focus on reducing child mortality and improving maternal health, both inextricably linked to nutrition and food security. The recent abrupt increase in food prices, in tandem with the current global economic crisis, threatens progress made in these areas and could prove a serious barrier to achievement of these goals.

WORKSHOP BACKGROUND

The Institute of Medicine (IOM), with funding from the Bill & Melinda Gates Foundation and the PepsiCo Foundation, held a workshop titled *Mitigating the Nutritional Impacts of the Global Food Price Crisis* on July 14–16, 2009, in Washington, DC, at the Kaiser Family Foundation's Barbara Jordan Conference Center. The workshop was a collaboration between the IOM Board on Global Health and the Food and Nutrition Board, in consultation with the Board on Agriculture and Natural Resources. Presenters were chosen by a planning committee to describe the dynamic technological, agricultural, and economic issues contributing to the food price increases of 2007 and 2008, and their impacts on health and nutrition in resource-poor regions. The planning committee quickly realized that it was impossible to ignore the compounding effects of the current global economic downturn on nutrition. Subject matter experts were invited to the workshop and asked to discuss these tandem crises, their impacts on nutrition, and opportunities to mitigate their negative nutritional effects. The primary objectives of this workshop were to:

- Set the stage for the deliberations by having an overview of the recent food price crisis and how it, in tandem with the current economic crisis, affects developing countries;
- Understand the pathways from the food price and economic crises to nutritional impact, including a discussion of existing evidence and vulnerable populations;
- Understand the range of country experiences with the food price and economic crises and their impact on food security and nutrition, as well as country-level responses to these crises;

---

[1] The MDGs were adopted in 2000 by the member nations of the United Nations and the world's major development institutions.

- Encourage a broad discussion of nutrition surveillance, including existing nutrition surveillance systems, their capacity to monitor food price fluctuations, and the gaps and needs for improved surveillance;
- Understand the landscape of the global nutrition field, those who work in it, and their respective roles and capacities to respond to the food price and economic crises; and
- Discuss what the U.S. government can and should do to help avoid future food crises and to mitigate the negative nutritional effects of those that cannot be avoided.

An IOM workshop illuminates scientific discussions to foster understanding among the public, academia, government, nongovernmental organizations, industry, and policy makers, but it does not make recommendations. A cornerstone of the approach is to air divergent views on sensitive and difficult issues in an atmosphere of respect and neutrality in order to encourage dialogue and strategic solutions.

The goal of the summary report is to present relevant lessons from the workshop, to outline a range of pivotal issues and their respective challenges and opportunities, and to offer potential responses as discussed by the workshop participants. The remarks in the workshop summary are the views of the individual presenters, panelists, or attendees and do not reflect a consensus of those attending or the planning committee. The planning committee's role was limited to planning the workshop, and the workshop summary has been prepared by the workshop rapporteurs as a factual summary of what occurred at the workshop.

The workshop summary is organized in chapters as a topic-by-topic description of the presentations and discussions as they occurred at the July 2009 workshop. The workshop agenda, as well as speaker information and a list of registrants, appears in the appendixes at the end of the workshop summary report.

The reader should be aware that the material presented here expresses the views and opinions of the individuals participating in the workshop and not the deliberations and conclusions of a formally constituted IOM consensus study committee. These proceedings summarize only what participants stated in the workshop and are not intended to be an exhaustive exploration of the subject matter and should not be perceived as a consensus of the participants, nor the views of the planning committee, the IOM, or its sponsors.

WELCOME FROM THE SPONSOR

Ellen Piwoz, Sc.D., M.H.S., Senior Program Officer
The Bill & Melinda Gates Foundation

The timing of this meeting is absolutely critical. Never before has the international nutrition community had the breadth and depth of data that exists today—showing the importance of nutrition to the lifelong health of children as well as that of their families, communities, and entire economies. Economic development and security require a healthy and fit population, and undernutrition undermines this process. In fact, it has been estimated that countries may lose 2–3 percent of their gross domestic product from deficiencies in such key nutrients as iron, iodine, and zinc (*Vitamin and Mineral Deficiency: A Global Progress Report*, 2004; Alderman, 2005). The increase in food prices has certainly brought this issue to the forefront, but it has also highlighted some of the challenges the international nutrition community is facing in a world where far too many people go to bed hungry and suffer from the consequences of malnutrition. It is hoped that this workshop can provide some tangible ways to respond, not only to the latest rise in food prices but also to ensure that nutrition is a component of every health and development strategy, in both donor and recipient countries alike, with the twin goals of improving nutrition while decreasing hunger and poverty.

The Bill & Melinda Gates Foundation's strategy for the nutrition program in global health is primarily focused on improving child nutrition from conception through 24 months; such improvements have a huge impact on the likelihood of a child's survival, good health, and later learning and earning potential. Given the importance that good nutrition plays in lifelong health and productivity, addressing nutrition early in life is one of the more cost-effective investments that can be made (Behrman et al., 2004; *Global Crises, Global Solutions*, 2004; Horton et al., 2008).

The fact that this very workshop is an initiative of the IOM's Board on Global Health, Food and Nutrition Board, and the Board on Agriculture and Natural Resources represents the complexity as well as the great opportunity facing the international nutrition community. No single agency or set of institutions has the sole mandate, or the resources, to coordinate a global response to the issue of undernutrition. But this cannot be an excuse for inaction. The recent announcement by the G8 to commit $20 billion toward food security is welcome and much needed. The international nutrition community must help to ensure that this strategy includes coordinated efforts to improve nutrition in addition to increasing food production.

Although the problems to be discussed over the next 3 days are very real and tragic for millions and millions around the world, there must also be a sense of optimism. Existing evidence shows that progress is possible on a large scale even in the poorest of countries. Today, the tools, solutions, and knowledge to address hunger and undernutrition exist. With the combination of commitment, capacity, and resources, successes have been demonstrated.

Over the past 5 years, the world has seen the introduction of new interventions—micronutrient powders and ready-to-use therapeutic foods—through the development of novel partnerships between researchers and the private and public sectors; these collaborations are beginning to have a broad impact. Countries from Asia to Africa to Latin America have been able to mount efforts

> **BOX 1-1**
> **The Millennium Development Goals (MDGs)**
>
> Goal 1: Eradicate extreme hunger and poverty.
> Goal 2: Achieve universal primary education.
> Goal 3: Promote gender equality and empower women.
> Goal 4: Reduce child mortality.
> Goal 5: Improve maternal health.
> Goal 6: Combat HIV/AIDS, malaria, and other diseases.
> Goal 7: Ensure environmental sustainability.
> Goal 8: Develop a global partnership for development.
>
> Three of the eight MDGs, adopted in 2000 by the member nations of the United Nations and the world's major development institutions, are closely linked to malnutrition in young children and in women: MDG 1—Eradicate extreme poverty and hunger, MDG 4—Reduce child mortality, and MDG 5—Improve maternal health. With a key milestone of the MDGs fast approaching in 2015, the goal to decrease hunger by 50 percent is far off track. The current economic environment makes achieving the goals even more difficult. Momentum to reduce overall poverty in the developing world is slowing, and, in particular, higher food prices have reversed the nearly two-decade trend in reducing hunger.
>
> SOURCE: UN, 2008.

to tackle such problems as vitamin A and iodine deficiencies. Many countries have been able to reduce the burden of underweight with a combination of direct nutritional interventions and other social and health investments. Child mortality rates are lower than ever before.

Yet only 6 years remain to achieve the Millennium Development Goals, Goal 1 of which is to "Eradicate extreme hunger and poverty" (Box 1-1). The past decade's progress is now threatened by the food and financial crises (UN, 2008). The international nutrition community must learn more about what is happening to people on the ground, which groups are being affected, and how these challenges can be overcome. An African proverb states, "If you want to go fast, go alone. If you want to go far, go together." This is the time to collectively mobilize knowledge, resources, and commitment to ensure that the most vulnerable populations of the world are not further condemned to a lifetime of poor health and suffering because of undernutrition caused today.

REFERENCES

Alderman, H. 2005. Linkages between poverty reduction strategies and child nutrition: An Asian perspective. *Economic and Political Weekly* 40:4837-4842.

Behrman, J. R., H. Aldermann, and J. Hoddinott. 2004. Hunger and malnutrition. *Copenhagen Consensus: Challenges and Opportunities* 58.

Global Crises, Global Solutions. 2004. Edited by B. Lomborg. Cambridge: Cambridge University Press.

Global Recession Increases Malnutrition for the Most Vulnerable People in Developing Countries. 2009. Rome: United Nations Standing Committee on Nutrition.

Hawtin, G. 2008. *CIAT's Response to the World Food Situation*. Cali, Columbia: CIAT.

Horton, S., H. Alderman, and J. A. Rivera. 2008. *Copenhagen consensus challenge paper: Hunger and malnutrition*. Copenhagen Consensus Center.

Ngongi, N. 2008. *Policy Implications of High Food Prices for Africa*. Washington, DC: International Food Policy Research Institute.

Sheeran, J. 2008. *High Global Food Prices: The Challenges and Opportunities*. Washington, DC: International Food Policy Research Institute.

UN. 2008. *United Nations Millennium Development Goals.* http://www.un.org/millenniumgoals/ (accessed October 27, 2009).

Vitamin and Mineral Deficiency: A Global Progress Report. 2004. Ottawa, Canada: The Micronutrient Initiative and The United Nations Children's Fund.

von Braun, J. 2008. Rising food prices: What should be done? *EuroChoices* 7(2):30-35.

2

The Dual Crises: Tandem Threats to Nutrition

Three of the eight Millennium Development Goals (MDGs), adopted in 2000 by the member states of the United Nations and the world's major development institutions, are closely linked to malnutrition in young children and in women: eradicate extreme poverty and hunger (MDG 1), reduce child mortality (MDG 4), and improve maternal health (MDG 5). A strong evidence base underpinning and motivating further investment in nutrition has emerged over the past 5 years. Specifically, a *Lancet* series, published in January 2008, showed that maternal and child undernutrition is the underlying cause of 3.5 million deaths annually, 35 percent of the disease burden in children younger than 5 years, and 11 percent of total global disability-adjusted life years (DALYs) (Black et al., 2008). With this evidence as a foundation, it seemed that if the will, the tools, and the technologies had all been mobilized, then real progress in nutrition could be made. Then the recent abrupt increases in global food prices, exacerbated by the current global economic downturn, began to threaten the hoped-for trajectory of progress. Between March 2007 and March 2008, price increases of 31 percent for corn, 74 percent for rice, 87 percent for soya, and 130 percent for wheat were documented (Hawtin, 2008). This chapter looks at the contributing factors and potential causes of the recent food price hikes and the current global economic downturn. As described by planning committee chair and session moderator, Reynaldo Martorell of Emory University, the following presentations helped to set the stage for the deliberations by having an overview of the recent food price crisis and how it, in tandem with the current economic crisis, affects developing countries.

THE RECENT AND CURRENT FOOD PRICE CRISIS AND FUTURE PERSPECTIVES

Per Pinstrup-Andersen, Ph.D.,
H.E. Babcock Professor of Food and Nutrition Policy
Cornell University

The "global food crisis" is often thought of as a crisis that came and went. In reality, this is a crisis that came but never went away. Until the beginning of this century, there had been a tremendous decrease in the real prices of food (the food price relative to other prices). Dramatic increases in food prices were seen during the past 2 to 4 years, up until the middle of 2008. Since then, prices began to fall, and they fell quite dramatically during the subsequent year, between the middle of 2008 until April or May of 2009. Yet the food crisis is far from over. Food prices are still considerably higher than they were 5 or 6 years ago. From a poverty and nutrition perspective, the *fluctuation* of the prices—or the price *volatility*—has the most significant impact (Figure 2-1). These fluctuations have led to an unstable nutritional environment, especially for the world's poor.

Causes of Food Price Fluctuations

A number of factors, specifically supply side, demand side, market, and public and private action, caused the prices of food to fluctuate. On the supply side these included adverse weather, rapidly falling prices (1974–2000), Organisation for Economic Co-operation and Development (OECD)[1] production subsidies, and limited agricultural investments. On the demand side, these factors included biofuel production, increased demand for meat and dairy products, increased feed demand, economic growth in middle-income countries, and economic recession. The market factors included low and falling stock levels, capital market transfers (capital invested into commodity futures), changing dollar value, and increasing oil and fertilizer prices. The public and private action factors included export bans and restrictions, panic buying, reduced import tariffs, price controls, rationing, food riots, hoarding, and media frenzy.

Supply Factors

A variety of supply factors affected price fluctuations. Changes in weather, especially short-term droughts, had a large impact on wheat producers in Australia, Ukraine, and parts of the United States. Droughts reduced production, which increased wheat prices. The rapidly falling prices that occurred after the

---

[1] The OECD is an international organization of 30 countries that accept the principles of representative democracy and free-market economy.

THE DUAL CRISES: TANDEM THREATS TO NUTRITION

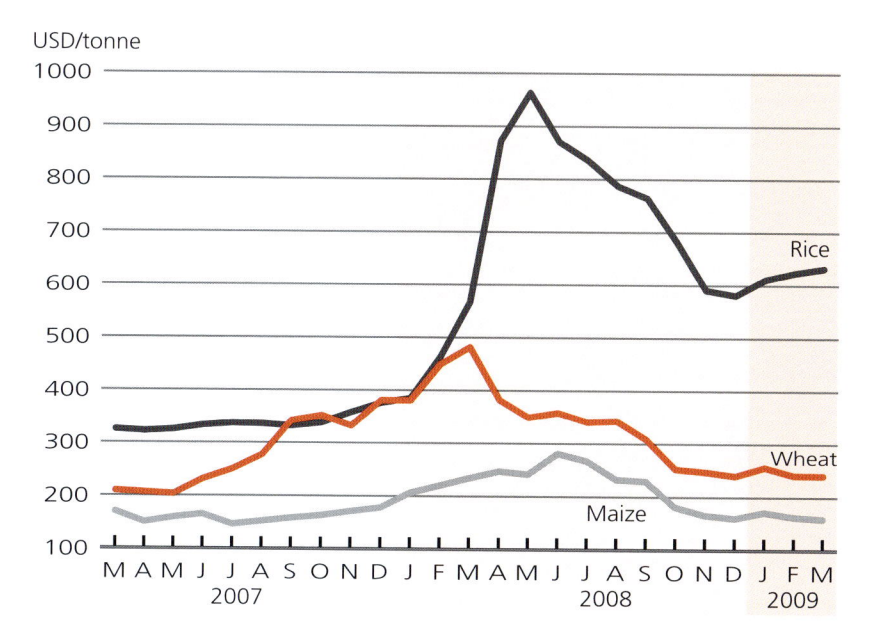

FIGURE 2-1 Selected international cereal prices.
NOTE: Prices refer to monthly average. For March 2009, 2 weeks average.
SOURCE: Food and Agriculture Organization of the United Nations, 2009b.

1974 price hike also affected the current food crisis. When prices continue to fall, little investment is made in rural areas and in agriculture. Therefore, agriculture productivity does not increase as fast as it should, production does not keep up with demand, and eventually prices begin to increase. In OECD countries, production-expanding subsidies were maintained, and more food was produced than the markets could bear at the higher prices. However, the products were dumped into the international market and even developing country markets, creating a very unfair advantage for the OECD countries and resulting in the depression of prices.

Demand Factors

When the prices of energy and oil rose, the United States, European Union, and a number of other countries turned to the use of food commodities to create biofuels. This increase in demand resulted in the pulling away of food commodities from the food market. Supply reduced, and food prices rose. Institutions' opinions as to how great an impact the use of biofuels had on the rapid price increase are usually in line with what the institution's economic interests are (some claim a 30 percent impact, others a 3 to 4 percent impact). Therefore, the extent to which biofuels increase food prices is unknown, but it is certain that biofuels had some effect.

While food prices were dropping, rapidly growing middle-income and high-income countries, such as China, increased their demands for meat and dairy. This put more pressure on the need for livestock feed, taking food away from the food market and causing prices to rise. Economic growth also contributed to an increase in the food demand. When the economic recession hit, however, prices came down again.

Market Factors

There were also various market factors contributing to the rapid increase in food prices. Stock levels, especially of grain, were very low from 1998 onwards. This added to concern that there would be price increases. Additionally, there was quite a bit of capital made available as part of the housing market problems and other financial crises-related issues in the United States and elsewhere. Much of that capital went into commodity futures. Oil prices were going up, food prices were beginning to go up, and people with money to invest decided commodity futures was a good place to put money. This phenomenon pushed up prices far beyond what the supply and demand called for. Later, when the capital was pulled out, there was a dramatic drop in prices, at least partly caused by this capital flowing out of the futures market.

Increasing oil prices pulled the prices of fertilizer and pesticides up as well, since oil is needed for the production of those commodities (in particular, nitrogen fertilizers). Fertilizer prices increased faster than the food prices and were much slower to come down after food prices began to decrease. Many farmers in Denmark, for example, were told in August and September 2008 that there was going to be a short supply of fertilizer for use in March and April 2009, and that they should buy fertilizer despite the high prices. Though farmers expected grain prices to remain high, they fell instead.

Yet another market factor for the food price increase was the depreciation of the U.S. dollar. Food prices are usually expressed in dollars; when the dollar became weak at the beginning of this century, food prices rose wherever they were expressed in dollars.

Public and Private Action Factors

Public and private behaviors also caused this rapid increase in food prices. Export bans and restrictions, particularly in the rice market, had an especially large impact. When wheat prices began to rise (due to drought, increase in biofuel production, and other reasons), the Indian government decided to stop exporting rice in order to protect itself and keep the rice price down, hoping to see a substitution between the higher-priced wheat and the lower-priced rice. This caused the rice price in the international market to increase dramatically. Other countries, such as Vietnam, Cambodia, and China, decided to follow India's lead in order to protect their own consumers. This is "bad neighbor policy," for the government sees its legitimacy and population as being threatened, and therefore has no concern about the World Trade Organization (WTO) and the rest of the world. Later, India and Vietnam continued to restrict exports, but gave political favors by allowing some select countries to purchase their rice. The United States, Thailand, and a few other countries decided to risk continued rice exporting; these "good neighbor policies" (though they were driven by self-interest) were the saving grace in the rice export market.

Reduced import tariffs also played a role in the rapid price increase. Countries put import tariffs on food to generate public revenue, but they began to take away the tariffs when international prices went up. The WTO did not have power over these policies so was unable to intervene. Countries faced with higher food prices put in place such things as price controls and food rationing. Households, and even governments, began to hoard food since they expected food prices to rise. In addition, there were food riots in up to 60 countries. These riots are not instigated by poor rural people whose children are dying from hunger and malnutrition, but rather by lower-income urban consumers who do not like paying more for food. The governments only respond to those who pose a threat to the government; therefore, no attention was paid to the people in the rural areas. This is why price controls, rationing, and input tariff reductions were put in place only in urban areas.

There was also a media frenzy surrounding the food crisis. When food prices went up, the food crisis was on the front page of virtually every newspaper in the world. However, as soon as food prices began to fall, the media stopped producing articles about the issues. This frenzy contributed to some of the panic and hoarding that happened after food prices began to rise. Even in the United States, some retail chains rationed rice for a short period of time.

Impact of Food Price Fluctuations

Factors that determine the impact of food prices at the household level include:

- Price transmission,
- Whether the household is a net buyer or net seller,
- Household budget share for food,
- Extent of value addition,
- Relative price changes among diet components,

- Relative price elasticities, and
- Risk management capacity.

The impact of food price fluctuations depends on the *price transmission*, or how the international market transmits prices to the country, household, and ultimately, to the individual. For example, consumers in India and China did not see much change in the rice price because the policies their governments had in place prevented consumers from being affected. These same policies, though, hurt others in the international community; for instance, they caused consumers in the Philippines to be greatly affected by rice prices. According to the Food and Agriculture Organization of the United Nations (FAO), although prices have come down, many low-income, developing countries are still feeling the effects of the food price spikes (Food and Agriculture Organization of the United Nations, 2009a). For example, in Kenya, the price for maize continued to increase as the international prices for maize decreased. Therefore, it is important to remember that there is *not* a direct link between global trade policies and the prices that poor people are faced with in developing countries.

Another factor that determines the impact of food prices on a household is whether the household is a net buyer or net seller. A large percentage of small farmers in developing countries are net buyers of food and therefore suffer short-term negative effects from food price increases. Organizations such as the Bill & Melinda Gates Foundation are trying to reach farmers with agricultural investments to help them become net sellers instead of net buyers. Once households overcome that threshold, they can benefit from higher food prices.

The impacts of the price increases also depend on the household budget share for food. Poor people spend 50 to 70 percent of their income on food. High-income people, on the other hand, spend 5 to 10 percent (International Bank for Reconstruction and Development, The World Bank, 2007). Furthermore, people in high-income countries spend a larger share of their consumer dollar on post-harvest additions to food, while poor country consumers' dollars go largely to pay for the food itself. Nutrition of the poor, then, is clearly much more affected by food price fluctuations. The relative price range among diet components can have very serious nutrition implications. If one particular food price goes up faster than another, households adjust to the change.

Finally, the effects of the food price fluctuations greatly depend on risk management capacity, which is very low among the poor. High-income people can utilize savings or cut back on luxury spending in times of crisis, whereas low-income people have to take much more drastic measures to adjust to the changes in price.

Policy Response to Food Price Increases: The Movement Toward Food Self-Sufficiency

Policy responses to the food price increases include expanding food production, moving toward self-sufficiency, and maintaining government legitimacy. Expanding food production has resulted in a renewed interest in national self-sufficiency and an increased interest in reserve stocks and the acquisition or control of land across borders. The move toward food self-sufficiency has built up global grain stocks, enhanced control over land in foreign countries, and increased the impetus toward a build-up of financial reserves. Efforts to maintain government legitimacy often emphasize such short-term measures as price controls, export bans, lifting import tariffs, rationing, and food distribution. Other efforts to maintain government legitimacy emphasize short-term transfers to the urban lower-middle class, which result in the continued neglect of the rural poor.

Governments can take a variety of positive steps, such as investing in rural areas, increasing productivity, funding agricultural research, and developing agricultural infrastructure. These allow farmers to produce more and increase their income so that they can finally escape poverty. The Bill & Melinda Gates Foundation has been supportive in helping farmers achieve this goal. These policy responses are positive ways that governments can gain more control over food price fluctuations.

Negative steps have also been taken. Governments are motivated to stay in power, which requires national support. This desire to maintain legitimacy when food prices began to increase led to price controls, export bans, reduced import tariffs, rationing, and food distribution. These governments emphasized short-term solutions that only addressed the unsatisfied and vocal urban lower-middle class, so as to diminish the threat this group posed to the government. These policies neglected the rural poor, who are continually ignored and malnourished.

Some governments expanded food production, showing a renewed interest in achieving national self-sufficiency. Food self-sufficiency is a dangerous idea that negatively affects the international community. Governments are rapidly building up reserve stocks to try to protect themselves from the untrustworthy international trade market. Additionally, some governments are trying to take control over land outside of their own country, often termed "land-grabbing." This might occur when middle-income countries move in to lower-income countries and attempt to gain control of large extensions of land and the crops that are produced on that land.

Ideally, countries need to focus their energies on building financial reserves in order to handle price volatilities. However, the economic crisis has made it difficult for countries to build such reserves.

Future Perspectives

What will happen in the future can be affected by a number of factors, including

- A significant supply response,
- Falling real food prices,
- Increasing price volatility,
- Strong links between oil and food prices,
- Continued urban bias in policy interventions,
- Return to government complacency, and
- $20 billion from the G8 that could dramatically increase food production and decrease poverty, hunger, and malnutrition.

When prices change, there is a significant supply response. For example, in response to price increases, farmers produce more food. Alternatively, when prices drop, farmers decrease the amount of food they produce. Lower-income farmers may not have the opportunity to expand and contract their production in response to the market, due to poor infrastructure, lack of available Green Revolution technologies, inability to get credit, and nonfunctioning markets. It is important to focus on the unique situation of those in low-income countries when trying to mitigate the impacts of food price increases.

Prices are eventually going to decrease, due to the large amount of food that will be released into the market over the next few years. Countries that have export bans will eventually have to open their doors for exports because they will be internally flooded with extremely high amounts of rice and grain stock.

An increase in price volatility in the future is highly likely because of climate change (more severe droughts and floods) as well as national efforts to become food self-sufficient. Attempts to be self-sufficient will cause increased food production where it should not necessarily be produced, while disincentivizing food production where it is most efficiently produced.

The strong links between oil and food prices may lead to a reconsideration of biofuel policies. Currently, without subsidies, it is not profitable to produce biofuels from corn. With higher food prices, there will likely be more investment in agricultural productivity. It would be ideal if developing country governments prioritize agricultural development and productivity increases. This is possible; the world is not running out of productive capacity. Yet, governments should not intervene in such a way that prices become skewed; that will affect not only economic efficiency, but also either producers, consumers, or both.

In the future, there will likely be a continued urban bias in policy interventions, owing to threats such as food riots that the urban poor pose against governments. The return to government complacency is a real danger to the health and nutrition of those most affected by the global food price crisis. Ideally, if the $20 billion that came out of the L'Aquila G8 summit in Italy is used effectively, there will be a dramatic increase in food production in Africa and a dramatic decrease in poverty, hunger, and malnutrition.

THE CURRENT GLOBAL ECONOMIC CRISIS AND FUTURE PERSPECTIVES

Hans Timmer, Director of the Development Prospects Group
The World Bank

The world is encountering a major global financial and economic crisis that has taken a heavy toll on private capital flows to developing countries. The financial crisis was sudden, global, and unprecedented. In September 2008, the world suddenly changed. Because of different policy reactions to this financial crisis, including aggressive stimuli by governments and severe reactions by monetary authorities, the World Bank predicts that the total crisis will not be as dire as the Great Depression of the early 1930s. International policy responses helped stabilize financial and commodity markets and stopped the "free fall" in trade and production. As a result of appropriate policy reactions, the worst of the collapse in the global economy is likely over. More worrisome, though, is the idea that this crisis—even if it lasts only a short time—is so deep that it will take several years to resolve. Coordination is needed to achieve true global recovery from the problems that many developing countries, in particular, are facing at the moment.

Impact on Developing Country Economies

The first mechanism through which developing countries were affected by this crisis was when, because of an increase in uncertainty, anxiety, and panic, financial markets saw a sudden drying up of credit. For example, the failure of Lehman Brothers had consequences that reached all over the world. As a result of the failure of big financial institutions in the United States, suddenly these institutions began to withdraw their overseas investments. Over the past several years, such investments had, to a large extent, gone to emerging economies. In fact, because there is strong growth potential in developing countries, there had been a boom in capital flows going to developing countries (mainly to the private sector) since the beginning of this decade. Such investments reached $1.2 trillion in 2007, dropping to around $700 billion in 2008 (International Bank for Reconstruction and Development, The World Bank, 2009a). The World Bank forecasts show that the capital flows going to developing countries will decrease further in 2009 to around $360 billion—a quarter of what resources were only 2 years ago (from 8 percent to 2 percent of gross domestic product [GDP] of all developing countries), significantly affecting developing countries (International Bank for Reconstruction and Development, The World Bank, 2009a) (Figure 2-2).

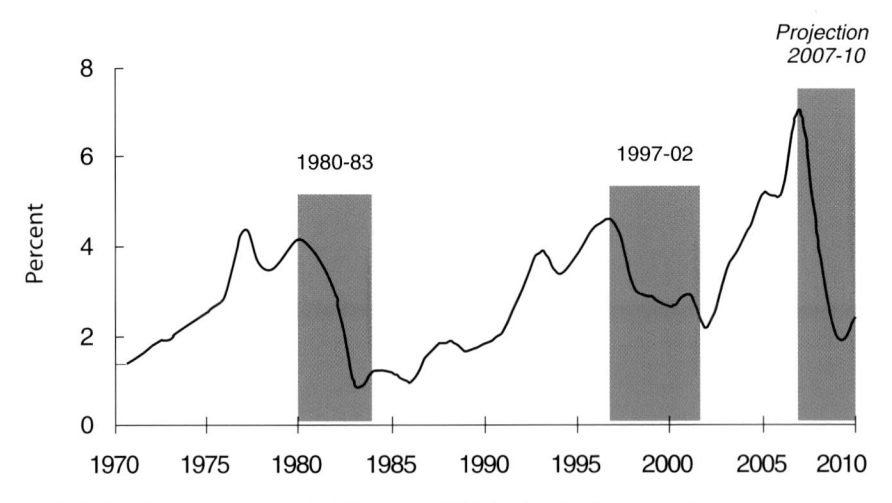

FIGURE 2-2 Net private capital flows per GDP in developing countries.
SOURCE: International Bank for Reconstruction and Development, The World Bank, 2009b.

In a macroeconomic sense, as a result of the crisis, all countries are facing financing needs this year. These needs are difficult to meet because of the collapse in financial flows, especially for central and eastern European countries that currently have large account deficits (they import more than they export and must repay money borrowed in previous years). In total, developing countries need some $1.1 trillion this year to repay former or finance current account deficits (International Bank for Reconstruction and Development, the World Bank, 2009b). Developing countries need support from international financial institutions to weather the storm of the financial crisis.

The World Bank gathers data from approximately 150 countries in the world and each month calculates the "heartbeat of the global economy"—or the growth of global industrial production. The recession after the burst of the dot-com bubble in 2000 and 2001 had global proportions. It is also notable that since that time, the world economy has experienced a period of very stable, strong growth. Developing countries especially performed very strongly over the past 5 years, but starting in September 2008, the points fall off the chart (Figure 2-3). Within 5 months, the world economy lost more than 15 percent of global industrial production (as much as was accumulated in the 4 years before September 2008). Global trade also experienced a similar, sudden decline.

In an attempt to determine the ingredients of a global recovery, it is important to understand the relation between this global decline in production and the global decline in trade, as well as the global decline in commodity prices. In the period of 5 months starting in September 2008, all commodity prices (including food

THE DUAL CRISES: TANDEM THREATS TO NUTRITION   23

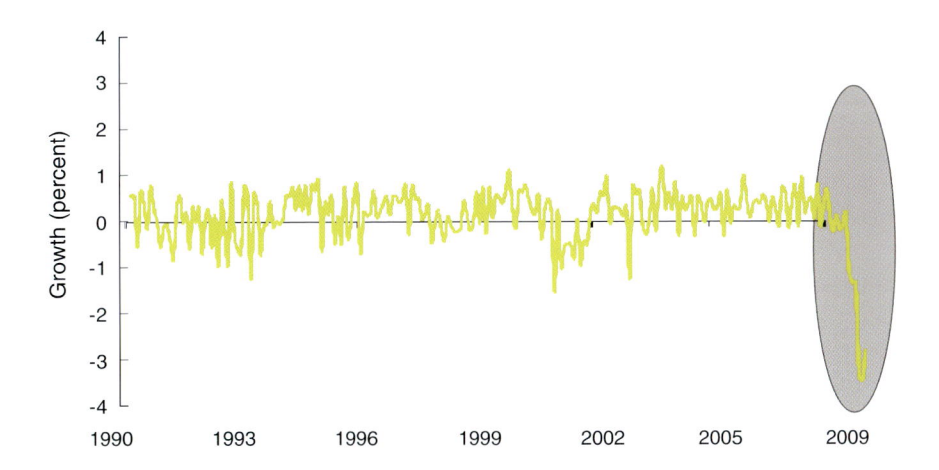

FIGURE 2-3 Rapid decline in global industrial production.
SOURCE: International Bank for Reconstruction and Development, The World Bank, 2009a.

and oil prices) fell between 40 and 50 percent (International Bank for Reconstruction and Development, The World Bank, 2009a, 2009b).

A common explanation of the global economic downturn assumes the world's dependence on the U.S. economy. According to this explanation, the crisis in the United States caused all developing countries that exported to the United States to feel the negative effects; as a result, the whole world economy is in a recession, and what is needed is large stimulus in the United States. The U.S. consumer needs to spend again, which will allow the rest of the world to recover. In the World Bank's view, this is the wrong story.

Instead, the World Bank's assessment is that, as a result of the financial crisis in the United States, all developing countries were affected in their domestic economies. Because of the withdrawal of capital flows and the global panic in financial markets, investors in developing countries stopped investing, and consumers in developing countries stopped purchasing. Within a month after September 2008, the car sales in India, South Africa, and even in China were 15–20 percent below the previous year's sales. This investment and consumer process is actually what impacted the global economy during the past 5 years.

Evidence for this explanation is seen in the import and export demand in the United States at the end of 2006. At that time, housing prices in the United States stopped increasing, and then started falling in the beginning of 2007. In the middle of 2007, housing prices fell at double-digit rates; the United States already had a financial crisis, but it was localized in the United States. As a result, U.S. consumers did not increase their consumption, and imports did not grow.

During this period, the developing world grew 8.1 percent—record growth in an historical perspective. Because of the financial crisis in the United States, U.S. exports declined. In developing countries between August 2008 and March 2009, the decline in imports was actually larger than the decline in exports. The cause of the financial crisis in the developing world, then, was not that they couldn't export to the U.S. market, but instead that their own domestic economies stopped functioning.

Thus, what is needed for a global recovery is recovery in the investment process in developing countries. Emphasis must be placed on coordinated stimuli that help developing countries restore their equilibrium so they can grow again.

Future Perspectives

By the end of 2009, a contraction of almost 4 percent of global GDP is expected. To a large extent this has been caused by the economic decline in high-income countries that face both domestic economic problems and reduced exports. On average, developing countries are still growing; a 1.2 percent growth is expected in 2009 (International Bank for Reconstruction and Development, The World Bank, 2009b). However, if India and China are removed from the list of developing countries, there would be a decline of 1.7 percent (International Bank for Reconstruction and Development, The World Bank, 2009b).

Because of the stabilization of the market and the myriad endogenous mechanisms that exist with a crisis like this, the World Bank predicts that the world will recover and even see 2 percent growth this year. Despite this percentage of recovery, at the end of 2010 there will still be less income being generated in the world than at the end of 2008. During these 2 years, the global economy will have lost income, which is unprecedented in the recorded history of the world.

In terms of the contribution to growth, the World Bank estimates that in 2010, more than one-half of growth will originate in developing countries, and half of that will be from China. This prediction is consistent with the performance of the world economy during the past 5 years. Developing countries contribute roughly one-quarter of the world economy, but over the past 5 to 7 years, their growth rate was on average 2.5 times the growth rates of high-income countries. In terms of contribution to growth, then, developing countries are already more important than high-income countries. It follows that the focus on where to stabilize the world economy should be in developing countries (Figure 2-4).

Despite the fact that recovery is predicted, the severity of the problems to be tackled—global underutilization of capacity, idle factories, and unemployment—may continue to increase. For example, unemployment will further increase in most countries next year and even in 2011.

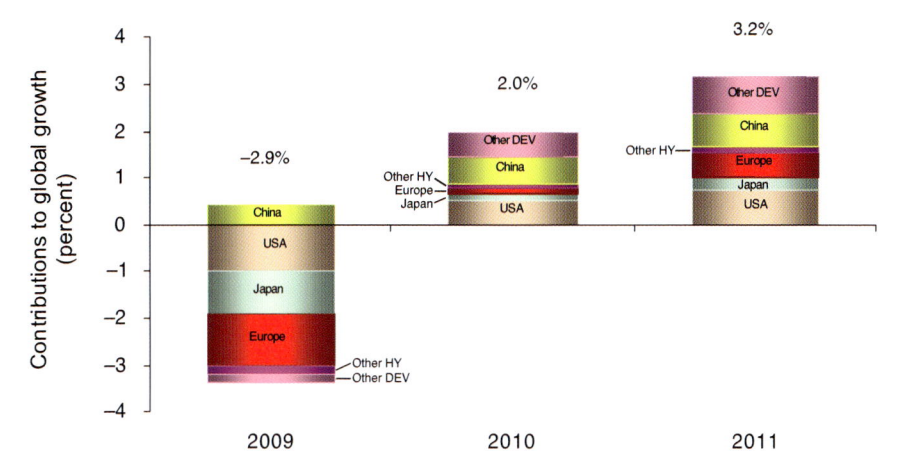

FIGURE 2-4 United States, China, and developing countries lead upturn.
NOTE: Other DEV = Other Developing; Other HY = Other High Income.
SOURCE: International Bank for Reconstruction and Development, The World Bank, 2009b.

Conclusions

First, while recent market indicators show stabilization, much remains to be done to place global finance on a stable footing. The global economic recession is a problem for governments that will require international support and coordination.

Second, protecting the growth prospects of developing countries serves the interest of rich countries as well. Major reforms in the developing world were seen during the 1990s that led to strong growth in many countries. It is not coincidental that the economic crisis did not start in countries with these emerging economies, but rather it began in high-income countries. There is still a great deal of potential in emerging economies, but they have been adversely affected by the crisis and need help to restore their economies.

Finally, the global nature of the crisis places a premium on policy coordination in order to prevent new crises and ensure balanced global growth. The current economic problems are occurring on such a large scale that no individual country can solve them on its own. International coordination should focus on two things: preventing new crises from developing and recovering growth in a balanced way that doesn't re-create the imbalances that contributed to the current crisis.

DISCUSSION

This discussion section encompasses the question-and-answer sessions that followed the presentations summarized in this chapter. Workshop participants' questions and comments have been consolidated under general headings.

Impact of the Crises on Poverty

The World Bank analyzed the impact of the economic downturn on poverty levels and determined that because of the crisis, up to 100 million people were pushed back into poverty (defined as living on less than $1.15 a day) or were not able to come out of severe poverty (International Bank for Reconstruction and Development, The World Bank, 2009). Although the poor are always the most vulnerable and it is always difficult for them to absorb shocks of any kind, the current economic crisis is actually one that does not necessarily impact the poor disproportionately. The economic shock decreases the value of assets and, therefore, affects the middle class in developing countries more than the extremely poor who have few assets. (Food prices, on the other hand, more than proportionately affect the poor.)

Impact of the Economic Crisis in Sub-Saharan Africa

The World Bank's first impression was that sub-Saharan African countries would be less affected by the financial crisis because they are not very integrated into global financial markets. It was later realized that the impact was large in many African countries because a large number of the investments in the mining sector were public–private partnerships. The private part of these partnerships ceased activity as private investors were hesitant, and they either postponed or cancelled their investments because of the financial crisis or the uncertainty in commodity prices.

More importantly, Africa is threatened by a *political* backlash. The reforms in Africa are much more recent than in many other parts of the developing world. Only recently did many sub-Saharan countries open their borders and invite foreign investors to come in. In Africa, this had a positive effect, and sub-Saharan Africa saw growth of 5–6 percent after decades of per capita income declines. The fear is that African national governments or opposition parties will now be reluctant to rely on international markets as a result of the crisis, and they will attempt to remain self-reliant.

Temporality of Food Price Spikes and Economic Downturn

The spike in food prices occurred in June and July 2008; the global economy dropped in the fall of 2008. Is there a temporal connection to be drawn between the two? In fact, what happened in the food markets is very consistent with what happened in other commodity markets, which is, in turn, very consistent with what happened in the global economy. The spike in food prices in mid-2008 can be explained by the fact that growth in the world economy was extremely strong. As a result, oil prices were very high. The link between oil prices and the food market is relatively new. What happened after September 2008 is exactly the reverse—a decline in global production. Demand for oil declined by 3.5 million barrels a day. Oil production decreased and oil prices came down, which brought down other prices.

Speculation

What role does speculation play in the food commodities market? Is there a way to regulate that type of commodity speculation when it comes to food staples? Should there be regulation of that kind? The World Bank does not believe that the financial markets were the cause of the swings in the food prices or in commodity prices in general. In addition, the World Bank does not believe that the financial markets were the cause of the sharp increase that was seen in the middle of 2008, and it is believed financial markets were not the cause of the downturn. What is very likely is that financial markets exacerbated some of the movement that was seen; it might be possible that the decline in food prices and commodity prices since the start of the crisis (that would have occurred anyway) occurred earlier and more quickly because of the working of financial markets.

It is not easy to regulate without damaging some of the advantages of financial markets—smoothing out behavior and giving either producers or consumers the opportunity to hedge the risks that they have to make. Even without financial markets there would have been large price swings. In fact, in several metal markets, the same kind of price behavior occurred though there were no financial instruments at work with those metals. It is easy to overestimate the importance of the financial markets.

Diet Diversity More Important Than Energy Consumption

There was a concern, voiced by Dr. Ricardo Uauy, that when food production is discussed, the conversation is only about cereals, and hunger should not be defined by energy or calorie intake alone; both diet diversity and the nutrient and micronutrient densities of diets are important in influencing nutritional health of populations. Food *quality* is what is important for nutritional well-being, and "hunger" should no longer be based on energy alone. The assumption that sufficient caloric intake for energy is all that matters has been misguided, and the global community is now paying the price of that. Diet diversity is what the international nutrition community needs to be concerned about. Statistics from FAO are still focused on calories and protein; a change in favor of empirical evidence on diet diversity and the intake of micronutrients would be an improvement.

Agricultural Foreign Direct Investment in Africa

The issue of foreign direct investment in Africa is controversial. While investing in agricultural productivity and the infrastructure needed to support it can contribute in positive ways, there are differing views on the process. Much of the effort that is taking place in this area is between two country governments, or between a private corporation outside the country and a government within the country. The risk of having that kind of negotiation is that small farmers who are currently on the land are being ignored; when these agreements are made, the farmers are often pushed off their land. If, instead, the agreements were made in negotiation with the farmers who are currently on the land, and if investment in rural infrastructure came along with such agreements, this type of agricultural foreign direct investment would have positive effects.

Trends in Development Assistance

Certain financial flows to developing countries are more resilient than other flows. For example, the flow of remittances (transfer of money by foreign workers to their home countries) is very resilient (and amounts this year to more than $300 billion to developing countries, which is almost as much as the private capital flows to developing countries and more than official aid to developing countries). The World Bank does expect a decline in that flow as a result of the labor market problems in countries of destination for migrants, but less so than is predicted in other flows to developing countries. Official aid, including philanthropies, is also a relatively stable source of growth. Financial flows coming from the World Bank are almost countercyclical in that the Bank is currently tripling its lending to developing countries.

Food Price Reasonableness

What constitutes "food price reasonableness," and how can it be measured? A reasonable food price is a function of productivity in agriculture and a function of people's purchasing power. There is no way to have a definition that declares the price of commodities without taking these variables into account; it is a very slippery concept. It is much easier to identify unreasonable prices than reasonable prices.

Sustainable Management of Natural Resources

Poverty, in many cases, creates unsustainable management of natural resources. If poverty can be reduced, particularly in rural areas, improved sustainable management of natural resources will follow. If rural people can be helped out of poverty, it is very possible that the sustainability in natural resource management can be increased. There is a misguided sentiment that if you increase productivity and agriculture is increased, natural resources are harmed. This is not necessarily so (Nkonya et al., 2008).

Investment in Human Capital

A significant proportion of recent economic growth has been the result of growth in developing countries. Some people in the nutrition and health field link that growth to investment in human capital. When discussing financial flows and eventual recovery strategies as part of the mitigation policy, is there any effort being made to preserve or continue the investment in human capital? Several workshop participants agreed that the international effort should not combat this crisis by merely spending money to create jobs in the short run. Instead, the focus should be on government interventions that help maintain or further increase productivity in preparation for the rebound. It is very important that government spending focuses on the long run, including investments in human capital.

Reliable Data for Early Warning in Africa

Obtaining reliable and current data from Africa is an ongoing challenge. This creates a big problem for timely response to crises. National governments' capacities to generate reliable data needs to be strengthened. The World Bank has a data group, and one of its missions is building capacity in developing countries, mainly in Africa. There is cooperation with governments and various institutes in improved data analysis, compilation, survey management, and organization.

Beyond direct data collection and analysis, there are indirect ways of trying to understand the impact of the crisis, such as watching the international markets and making comparisons with other countries. A degree of creativity may also be needed when working with early warning systems. Some workshop participants argued that, in some situations, although the hard data do not exist, action can and should be taken. It is not always prudent to wait until perfect data collection exists in all countries.

The World Bank Response

In responding to the food price crisis, the World Bank focused on three areas, starting with safety nets. A safety net refers to the institutions through which targeted transfers to those who need help most can be facilitated. The focus is much more on institutions than on money.

The second area the World Bank has focused on is investment, especially in Africa. If a very long-term view is taken, the problem is not that the world is running out of food. There is, instead, a continuous increase in the production of food and, in particular, meat. The problem is that there are regions where the demand is relatively high and the supply growth is lacking. Africa is the perfect example of that. Decades ago, Africa was a food exporter; it is now one of the largest food importers. A solution to the global food crisis, then, is to increase productivity in Africa.

The third area of focus for the World Bank was an attempt to stabilize prices through international government intervention in stock keeping and regulation of financial markets. After much debate, though, the Bank decided that such efforts have always failed, been ineffective, and turned into price subsidies and price management.

Effects of the Crisis on Geopolitics

Every major crisis changes the world in some lasting way. After this crisis, the world will be never the same because the role of developing countries will be much larger than it was before. Additionally, the very dominant position of the United States in financial markets and international policy making will be diminished somewhat, and the role of other countries—especially in Asia—will increase. The G20 is already more important than the G8 in many policy discussions, part of a growing trend toward increasing the voice of developing countries.

REFERENCES

Black, R. E., L. H. Allen, Z. A. Bhutta, L. E. Caulfield, M. de Onis, M. Ezzati, C. Mathers, and J. Rivera. 2008. Maternal and child undernutrition: global and regional exposures and health consequences. *Lancet* 371(9608):243-260.

Food and Agriculture Organization of the United Nations. 2009a. "Global Information and Early Warning System." Retrieved September 3, 2009, from http://www.fao.org/giews/english/index.htm.

———. 2009b. "GIEWS International Cereal Prices." Retrieved September 3, 2009, from http://www.fao.org/giews/english/ewi/cerealprice/2.htm.

Hawtin, G. 2008. *CIAT's Response to the World Food Situation*. Cali, Columbia: CIAT.

International Bank for Reconstruction and Development, The World Bank. 2007. *World Development Report 2008: Agriculture for Development*. Washington, DC: The World Bank.

———. 2009a. *Global Development Finance 2009: Charting a Global Recovery*. Washington, DC: The World Bank.

———. 2009b. *Global Economic Monitor*. Washington, DC: The World Bank.

New Challenges Emerge at Millennium Development Goals' Half-Way Point. 2009. Retrieved November 3, 2009, from http://web.worldbank.org/WBSITE/EXTERNAL/NEWS/0,,contentMDK:21914817~pagePK:64257043~piPK:437376~theSitePK:4607,00.html.

Nkonya, E., et al. 2008. *Linkages between Land Management, Land Degradation, and Poverty in Sub-Saharan Africa: The Case of Uganda*. Washington, DC: International Food Policy Research Institute.

3

Impacts on Nutrition

The global poor are usually the hardest hit by food price increases and economic strife. At the household level in developing countries, poor consumers spend 50–70 percent of their budget on food (von Braun et al., 2008), so their capacity to respond to higher food prices (or reduced incomes) is limited, and often forces households to make difficult choices that adversely affect women and children. This chapter highlights the theoretical pathways and evidence base around how the dual crises are affecting nutrition. As described by moderator Isatou Jallow of the World Food Programme, the following presentations helped workshop participants to understand the pathways from the food and economic crises to nutritional impact, including a discussion of existing evidence and vulnerable populations.

CONCEPTUAL PRESENTATION ON PATHWAYS TO NUTRITIONAL IMPACT

Ricardo Uauy, Ph.D., Professor, Public Health Nutrition and Pediatrics
London School of Hygiene and Tropical Medicine;
Institute of Nutrition INTA, University of Chile

The pictures below illustrate the substantial difference in food expenditures of families in different areas of the world. This demonstrates how income directly affects the ability to maintain good nutrition through food consumption (Figures 3-1 to 3-6).[1]

---
[1] All monetary amounts are in U.S. dollars.

A Survey of Food Expenditures Around the Globe

In Germany, families spend about $500 per week to feed a family of four. There is much variety, including a great deal of processed foods, although fresh fruits and vegetables are also prominent in the household (Figure 3-1).

In Cuernavaca, Mexico, families spend about $189 per week. Here fruits and vegetables are abundant, although processed foods and sweetened beverages figure prominently (Figure 3-2). In this situation, if it is necessary to make do with a reduced income, it is possible to decrease food quantity without necessarily sacrificing the food quality. Ironically, a reduced income might cause the family to cut out the unnecessary processed foods and soft drinks, which would improve this family's nutritional status.

In Cairo, a family spends $69 dollars per week on food. This amount of weekly expenditure in Egypt still enables a fairly varied diet (Figure 3-3).

In Quito, Ecuador, however, families spend about $32 on food, and sacks of cereals, wheat, and some legumes are featured prominently (Figure 3-4). Less fruits and vegetables are seen as compared to the previous families' diets. In this scenario where there is less variety, if some foods are eliminated from the picture, the family's consumption will suffer in nutritional quality.

FIGURE 3-1 One week of food for the Melander family in Germany.
SOURCE: Menzel and D'Aluisio, 2005.

IMPACTS ON NUTRITION 33

FIGURE 3-2 One week of food for the Casales family in Mexico.
SOURCE: Menzel and D'Aluisio, 2005.

FIGURE 3-3 One week of food for the Ahmed family in Egypt.
SOURCE: Menzel and D'Aluisio, 2005.

FIGURE 3-4 One week of food for the Ayme family in Ecuador.
SOURCE: Menzel and D'Aluisio, 2005.

A family in Bhutan can only afford $5 per week for its food. There is less food overall, and it is basically plant foods, including fresh fruits and vegetables. There are less animal foods, as grains figure prominently (Figure 3-5).

A family in Chad spends only $1 on food each week. The essence of their meager diet is cereals and some legumes, and almost exclusively features plant foods (Figure 3-6).

Diet Quality Suffers in Times of Crisis

These pictures demonstrate what foods people buy with the amount of money they have to spend on food each week. While these photos convey the present status of these populations, they suggest what people might stop buying if they had less money—during a food crisis, for example. In crisis situations, families preserve diets based on less expensive foods. If their income is sharply reduced, families do away with animal foods and nonstaple foods. They eat less meat, less dairy, less processed foods, less vegetables, and less fruits; they are predominately dependent on cereals, fats, and oils. They find ways to get adequate energy at a very low price, but may forego appropriately nutritious foods, which results in poor quality diets that are inadequate in terms of micronutrients (unless these sources are fortified, such as the fortified cereal and oil provided by the UN World

IMPACTS ON NUTRITION 35

FIGURE 3-5 One week of food for the Namgay family in Bhutan.
SOURCE: Menzel and D'Aluisio, 2005.

FIGURE 3-6 One week of food for the Aboubakar family in Chad.
SOURCE: Menzel and D'Aluisio, 2005.

Food Programme [WFP]). Continued access to energy-dense, micronutrient-poor diets in urban setting leads to increased risk of obesity, setting the stage for diet-related chronic diseases of adults (Victora et al., 2008). Micronutrients need to be preserved in diets, even during times of crisis.

Poor-Quality Diets Increase Morbidity and Mortality

Income has a major effect on the access that various populations have to a nutritious variety of foods, specifically animal foods, fruits, and vegetables. Income not only affects what people buy and what they eat, but it also directly affects the health of children, as well as their cognitive and immune functions. Poor nutrition leads to increased infection, which, in turn, compromises nutrition. Breast milk is a protective factor during the crucial first year of life and beyond, both in terms of its capacity to provide food energy and protein, but also as an immune enhancer and a protective factor against infections (Black et al., 2008). In fact, breast-feeding promotion is the most cost-effective means of saving 1 million out of the 3.5 million children younger than 5 years that are presently dying from all causes related to nutrition. Exclusive breast-feeding is recommended for the first 6 months of life. However, suboptimum breast-feeding, especially non-exclusive breast-feeding in the first 6 months of life, is estimated to be responsible for 1.4 million deaths and 10 percent of the disease burden in children younger than 5 years (Black et al., 2008). Among children living in the 42 countries with 90 percent of child deaths, a group of effective nutrition interventions including breast-feeding, complementary feeding, and vitamin A and zinc supplementation could save about 2.4 million children each year—25 percent of total deaths (Jones et al., 2003).

Poor Nutrition Also Has Lasting Economic Impacts

Deficiencies in diet quality during the ongoing food crisis have the greatest impact on women of reproductive age and children under 3 years, especially children under 18 months. At the same time, families with fewer resources will use less money for education, housing, and medical care, likely compounding the effects of the food price crisis for the most vulnerable groups.

A recent examination by the Economic Commission for Latin America showed that only 10 percent of the cost of hunger in Latin America was related to health or poor educational performance. Instead, almost 80 percent of the negative consequences of hunger were linked to reduced productivity throughout the life course. In this way, economic growth will be restricted by the productivity of those children who are suffering the negative impacts of the global food crisis, and that will have a long-term, transformational effect upon society's development. In addition to the short-term nutritional effects of the food crisis, then, there may also be long-term effects that need to be considered.

Conclusion

The following conclusions were drawn about the food crisis and its impact on nutrition:

- Demand for food staples that provide energy (rice, wheat, maize, sugar) is less affected by prices and income than nonstaple foods (meat, dairy, pulses, fruits, and vegetables).
- Critical essential micronutrients (vitamins and minerals) are concentrated in nonstaple foods. The consumption of these more expensive, less affordable foods is very price sensitive.
- Present global malnutrition is characterized by poor intake of vitamins and minerals, resulting in high prevalence rates of micronutrient deficiencies.
- Modest decreases in current intakes of vitamins and minerals will increase prevalence of micronutrient deficiencies, affecting the short- and long-term nutrition and public health of the poor.
- The challenge is not only to prevent a reduction in the quantity of food, but also to preserve the quality of the food consumed. The solution to hunger and malnutrition is not achieved by providing only food energy in sufficient amounts; the food should also be of adequate micronutrient content and quality.
- Unfortunately, micronutrient-rich foods are normally the more expensive foods. The international nutrition community must make a stronger effort at fortification and biofortification in order to make micronutrients available for vulnerable groups, especially children and women of reproductive age, and especially during times of crisis.
- The food price crisis is already having a significant impact on the world's ability to reach the first UN Millennium Development Goal (MDG). The number of food-insecure people has increased; thus, the goal of halving the number of hungry people by 2015 will certainly not be met. Moreover, malnutrition in its various forms will restrict the international community's capacity to meet all other MDGs.

EXISTING EVIDENCE OF NUTRITIONAL IMPACTS

Francesco Branca, M.D., Ph.D., Director,
Department of Nutrition for Health and Development
World Health Organization

The Global Impact

For the first time in human history, more than 1 billion people are undernourished worldwide. The Food and Agriculture Organization of the United Nations (FAO) estimate that 1.02 billion people worldwide—one-sixth of humanity—are undernourished (Food and Agriculture Organization of the United Nations, 2009). This is about 100 million more than last year. Unless substantial and sustained remedial actions are taken, the Millennium Development Goal and World Food Summit target of reducing the number of undernourished people by half by 2015 will not be reached. The latest estimates of the FAO show a significant deterioration of the already disappointing trend witnessed over the past 10 years. The spike in numbers of undernourished in 2009 underlines the urgency to tackle its root causes swiftly and effectively.

The current global economic slowdown, which follows and partly overlaps the food and fuel crisis, is at the core of the sharp increase in numbers of undernourished in the world. It has reduced incomes and employment opportunities of the poor and significantly lowered their access to food. The increase in undernutrition is not a result of limited international food supplies. Recent figures of the FAO Food Outlook indicate strong world cereal production in 2009.

With lower incomes, the poor are less able to purchase food, especially where prices in domestic markets are still high. While world food prices have retreated from their mid-2008 highs, they are still high by historical standards. Also, prices have been slower to fall locally in many developing countries.

At the end of 2008, domestic staple foods still cost on average 24 percent more in real terms than 2 years earlier, a finding that was true across a range of important foodstuffs (Food and Agriculture Organization of the United Nations, 2009). The combination of lower incomes caused by the economic crisis and persisting high food prices has been devastating for the world's most vulnerable populations.

The declining trend in the rate of undernutrition in developing countries since 1990 was reversed in 2008, largely because of escalating food prices (Food and Agriculture Organization of the United Nations, 2009). As seen in Figure 3-7, 642 million of the world's undernourished are in Asia and the Pacific, with a majority in developing countries; almost 300 million are in sub-Saharan Africa. Both these regions have seen a 10 percent increase in numbers of undernourished from 2008 to 2009 (Food and Agriculture Organization of the United Nations, 2009).

Because these data are from 2007, they do not answer the question as to

IMPACTS ON NUTRITION

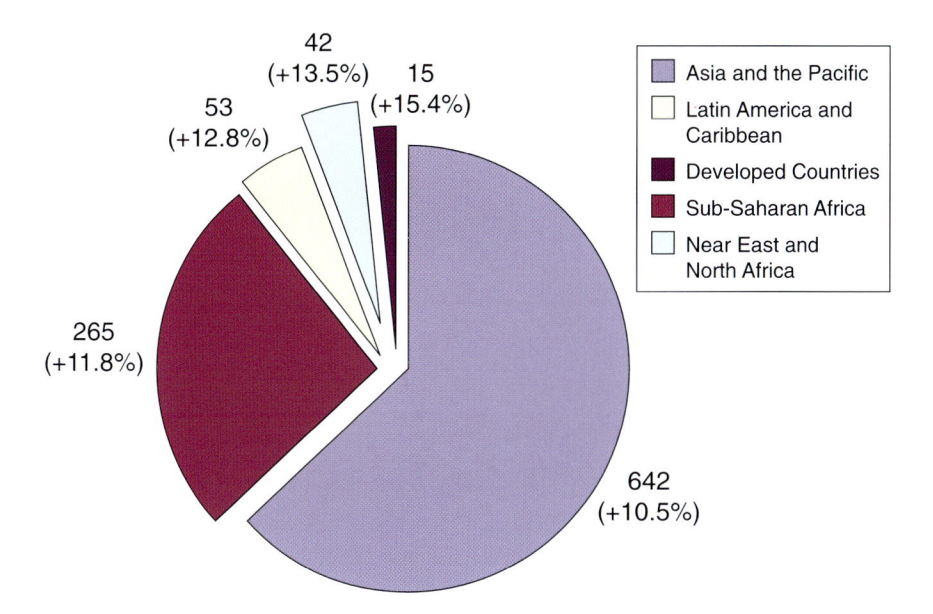

FIGURE 3-7 Estimated regional distribution of undernourished in 2009 (in millions) and increase from 2008 levels (in percentage).
SOURCE: Food and Agriculture Organization of the United Nations, 2009.

whether the food price crisis produced these results. It is likely that a combination of events caused the reversal in the declining trend of numbers of undernourished, including the HIV epidemic and major human-made disasters. How, then, can the impact of the food price crisis be measured? A number of organizations such as Action Against Hunger and Save the Children have been looking at data in newspapers and existing literature; other organizations such as the World Health Organization (WHO) and The United Nations Children's Fund (UNICEF) have been actually collecting data, but the data are sparse and anecdotal. It is very difficult to have a comprehensive, systematic picture of the impact of the food price crisis on the nutritional status of vulnerable groups.

Lessons have been learned from previous crises. The 1994 devaluation of currency in the Congo led to an increase in wasting among mothers, more babies born with low birth weight, and a greater level of stunting and wasting among children. A few years later in Indonesia, the financial crisis led to increased wasting in mothers and higher prevalence of anemia in mothers and children, although strangely no increase in childhood malnutrition. The Dutch famine of 1944 showed that the negative experiences of mothers during the famine had significant consequences on their children, including not only low birth weights, but also increased risks of noncommunicable diseases like obesity, mental disorders, behavioral problems, increased blood pressure, and coronary heart disease. This knowledge should cause great concern about what will happen to those who are now experiencing food shortages.

What Nutritional Impacts Are Expected?

Not everyone is affected equally by high food prices. It is expected that households that are net food buyers would lose with rising food prices, while net food-selling households stand to gain. Even within households, however, individual members are likely to be affected by food crises in different ways, with the nutritionally vulnerable members—women of childbearing age and young children—most at risk.

Households that are net consumers in rural and urban areas are most likely to face the negative effects of high food prices, and households' resilience to these increases will depend on their available coping strategies. Typically, poor households have limited strategies in which to cope with such shocks. Reducing household expenditure in other areas may mean a reduction in spending on education and health service utilization, therefore affecting child well-being. The impact on girls' education and nutrition in certain areas of the world may be particularly severe, due to strong cultural preferences toward sons (girls are less likely to be kept in school and may be fed differently than boys).

A decrease in purchasing power causes people to purchase fewer nutrient-dense foods, such as animal-source foods (meat, poultry, eggs, fish, and milk), fruits, and vegetables. When the "savings" brought about by this coping strategy is not enough, they may also reduce expenditure on basic foods, such as sugar, oil, and salt, as well as staples. In this way, the intake of specific nutrients, in particular *micro*nutrients, is reduced before energy intake is reduced.

The Impact of the Food Crisis: Evidence from Case Studies

To evaluate exactly what the effect of the food crisis is in various countries, their particular resilience and capacity to cope must be understood. Such capacity is very different in various countries and even within households. The following case studies, from Tajikistan, Ethiopia, Cambodia, and Sierra Leone, paint a picture of how four countries have reacted to the crisis.

Tajikistan

In Tajikistan, 91 percent of households are net food buyers. The shock of food price increases in 2008 caused a severe increase in undernourished in this country. In 2009, the majority of food-insecure households improved to the point where they are only moderately food insecure. For these households, access to food has improved, mostly thanks to the return of migrants, the transfer of remittances before the winter months, and the stable (but still high) food prices. On the other hand, these households remain vulnerable to shocks as most have not built sustainable assets and continue to use such negative coping strategies as skipping entire days without eating or eating "famine foods" such as seeds. This difficult and precarious economic situation is forcing households to contract new debts mostly to buy animal feed. One-half of the households surveyed have taken on new debts, and one-third of them will take longer than 2 months to pay them back (International Bank for Reconstruction and Development, The World Bank, 2009).

Ethiopia

In Ethiopia, during the 2008 food price hikes, the cost of maize increased by more than 75 percent, cabbage by more than 66 percent, and haricot beans by more than 20 percent (Action Contre la Faim, 2009). High international prices of cereals and inflated oil prices in the region had some effect, but drought and conflict remain the most significant factor behind the price rises of locally produced foods. Consumption of high-quality, micronutrient-rich foods was reduced while staple food consumption remained largely the same. In Ethiopia, it is likely that staple foods will continue to be replaced by cheaper, lower-quality staples: maize will be replaced by *kocho*, which has a lower vitamin A and protein density than maize (Abebe and Singburaudom, 2006). At the national level no increase in malnutrition rates were seen. However, survey data from 3 districts indicate that rates of malnutrition and under-five mortality increased in late 2007 and early 2008, corresponding with high food prices. No change in malnutrition rates was seen at the national level, indicating that country-level data are too imprecise to inform policy making and that surveillance data from the local level is needed (Action Contre la Faim, 2009).

Cambodia

Cambodia is classified as chronically food insecure as defined by the Integrated Food Security Phase Classification. All of the indicators are very close to crisis levels, and there are signs of negative change. The urban poor may have been more affected by rising food prices than the rest of the country; in urban areas, consumption of nearly all food groups (12 of 14) has decreased, and mean consumption has dropped from 5.4 to 4.8 food groups. Consumption of meat and fish has dropped 14 percent. Perhaps the most alarming finding is that the percentage of wasting among the urban poor has risen from 9.6 percent in 2005 to nearly 16 percent in 2008 (Food and Agriculture Organization of the United Nations, 2008).

Sierra Leone

Since 2002, Sierra Leone has been recovering from civil war. Food prices are intensely political, and undernutrition remains a threat to long-term security. Malnutrition remains very high, with 27 percent of children under 5 years reported as underweight. High inflation (more than 10 percent), stunting rates (37 percent), and poverty levels (between 65–75 percent) prompted FAO to rank Sierra Leone sixth in its assessment of nations most vulnerable to global price rises (Action Contre la Faim, 2009).

From January to March 2008, rice prices increased by 64 percent, and fuel prices increased by about 15 percent between January and May. The price rises corresponded with the beginning of the annual "hunger season," when families rely more on imported rice because local produce is more expensive and in short supply before the harvest. A survey was conducted in Freetown, home to more than 760,000 people, of whom 60 percent are under 25 years old and 97 percent rely primarily on the market for food. Areas experiencing the greatest increase in prices were also the areas where people most decreased the quantity of rice eaten per day. Meat consumption was most radically affected, and 43 percent of respondents reported they no longer consumed meat (Action Contre la Faim, 2009).

Conclusions

A number of conclusions and recommendations can be offered. The first conclusion is that the available data are poor. The data are often sparse and often anecdotal, although good monitoring systems on prices of commodities and on expected numbers of undernourished exist. However, nutritional surveillance is not performed systematically.

One recommendation to improve this situation emphasizes better surveillance. To achieve this, protocol standardization and technical and financial resources are needed. The best methodology for improved surveillance needs to be determined; for example, can national surveys and hot spot monitoring be combined? In addition, the nature of the data must be broadened to include food consumption, nutritional status, health outcomes, socioeconomic and policy context, growth velocity, and micronutrient data.

The second conclusion is that nutritional impact varies by population and is possibly linked to the extent of stress and resilience. Variations in dietary intake and coping mechanisms in response to an increase in food prices include reduction of nutrient-dense foods, reduction of meals, reduction of portions, and use of "famine foods."

The recommendations for dealing with the varying nutritional impacts on the population are to detect hidden nutritional effects and to identify and monitor vulnerable populations. To achieve this will require documenting nutritional impacts in population subgroups; identifying and monitoring micronutrient deficiencies in vulnerable groups; tracking such long-term effects as low birth weight and chronic diseases; following population groups, such as the elderly; monitoring vulnerable households; and evaluating the coverage and impact of nutrition interventions for targeted populations.

ARE THE URBAN POOR PARTICULARLY VULNERABLE?

Marie Ruel, Ph.D., Director, Poverty, Health, and Nutrition Division
International Food Policy Research Institute

Poverty Better Indicator of Food Crisis Impacts Than Geography

In the next quarter century, the population explosion that characterized much of the previous century will be replaced by another dramatic transformation: urban population growth on an unprecedented scale. About half of the world already lives in cities, and by 2025 almost 50 percent of Africans and Asians and 80 percent of Latin Americans will live in urban areas (UN, 2004).

Urban dwellers have limited access to agriculture and to land, so they must purchase most of their food; they are net buyers, which means they purchase more of their calories than they produce. For this reason, they need to have access to income. Urban people have to work in order to get cash to buy food. Because they need employment, very often women work outside the home in addition to men.

Are the urban poor truly particularly vulnerable to the food price crisis and its impacts? The evidence regarding whether or not the urban poor really are suffering the most is largely anecdotal. It often comes from small context-specific and nonstatistically representative studies. The media has declared there needs to be a focus on the urban poor, and the urban poor also have made themselves heard with food riots. The hypothesis is that the food price crisis affects the urban poor disproportionately. Intuitively, it seems this must be true because they purchase most of their food; so as food prices rise, the urban poor must be unduly affected. With the arrival of the current global economic crisis, it is presumed that the phenomenon has intensified the vulnerability of the urban poor because it has reduced employment opportunities. At this point, no documented information regarding the effects of the economic crisis on urban populations has been found.

Figure 3-8 is a conceptual framework on the determinants of child nutrition. On the left side are listed the different factors affecting household access to food: income, prices, production, and transfers and remittances. The relative importance of these factors differs in urban and rural areas because in urban areas people have less access to production and therefore they need income, or jobs to generate income; they also need reasonable prices to have adequate access to food. If none of these factors exist, urban dwellers use transfers, remittances,

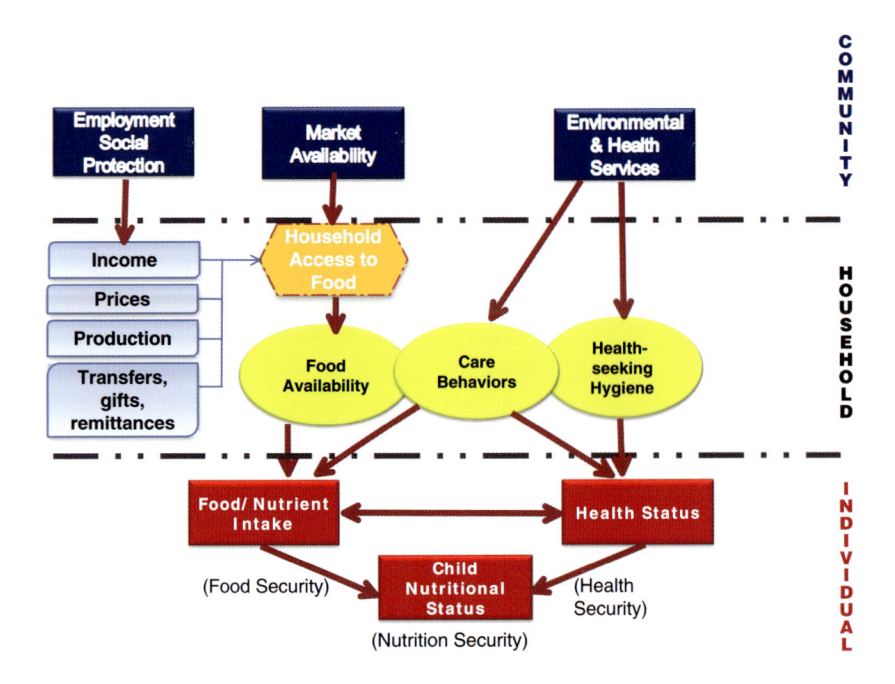

FIGURE 3-8 Determinants of food, nutrition, and health security in urban areas.

or other social protection mechanisms. The food price crisis is affecting these access factors, and it is affecting overall food availability at the household level. The response of households and their coping strategies—how resources are reallocated in terms of who works, how many hours they work, whether children are taken out of school to work, who eats what—is where households can make some adjustments. They can change such behaviors as their use of health services or the allocation of resources within the household and make decisions about who works, for how many hours, and where.

It is widely accepted that urban populations are net buyers of staple foods; what is often not discussed, however, is that there is also a very large proportion of *rural* dwellers who are also net buyers of staple foods. Table 3-1 shows nine different countries; overall, 96 percent of the urban populations are net buyers versus 74 percent of the rural dwellers. Among the poor, the rural poor have the highest percentage—88 percent are net buyers. In this light, the rural poor *and* the urban poor are all net buyers, meaning *poverty* is probably more important than geography in determining the effects of the food price crisis.

A special 2008 issue of *Agricultural Economics* also found that the poorest are the ones who suffer the most from the food price crisis, irrespective of

TABLE 3-1 Net Buyers of Staple Foods: 96% of Urban and 74% of Rural Dwellers

	All Households			Poor Households		
	Urban (Percentage)	Rural	All	Urban	Rural	All
Albania, 2005	99.1	67.6	82.9	*	*	*
Bangladesh, 2000	95.9	72.0	76.8	95.5	83.4	84.2
Ghana, 1998	92.0	72.0	79.3	*	69.1	*
Guatemala, 2000	97.5	86.4	91.2	98.3	82.2	83.1
Malawi, 2004	96.6	92.8	93.3	99.0	94.8	95.0
Nicaragua, 2001	97.9	78.5	90.4	93.8	73.0	79.0
Pakistan, 2001	97.9	78.5	84.1	96.4	83.1	85.4
Tajikistan, 2003	99.4	87.0	91.2	97.1	76.6	81.4
Vietnam, 1998	91.1	32.1	46.3	100.0	40.6	41.2
Unweighted average	96.4	74.1	81.7	97.2	87.9	78.5

* No data available.

country, region, and geographic area. The journal also found that female-headed households suffer disproportionately. The most vulnerable groups were found to be the poorest, the urban, those who do not own land, the nonfarmers, the larger households, the less educated, the less well-served by infrastructure, and those who live in a rural area and have limited access to land and modern agricultural inputs—essentially people who are poor or ultra-poor and need money to purchase food (Headey and Fan, 2008).

Coping Strategies

The impact of the global food crisis on food security will depend a great deal on the food-related coping strategies that affected households adopt. There are strategies that revolve around food, such as switching to cheaper, less preferred, lower-quality foods; buying less food, skipping meals, and reducing food intake; decreasing intake of nonstaple foods; eating out and increasing consumption of street foods; using different ingredients and cooking methods; and modifying the allocation of resources within the household.

There are also nonfood coping strategies such as boosting agricultural production and even returning to rural areas if the household owns land; augmenting income through child labor and women working outside the home; increasing the number of hours worked; taking children out of school; reducing spending on nonfood (health, education); and other desperate measures such as sending children to live with relatives or putting them up for adoption. The most detrimental of these coping strategies are those that include deinvesting in children, as noted above, as this will have long-term consequences on those children's development and future earning potential.

Conclusions and Lessons Learned

The following conclusions and lessons learned were offered:

- The urban poor are clearly vulnerable to the food price crisis and will most likely be extremely vulnerable to the financial crisis as well.
- Landless rural poor (who were vulnerable to start with) are net buyers and are extremely vulnerable; the dichotomy of the urban poor versus the rural poor may not be necessary except when you think of delivery mechanisms for interventions.
- The magnitude and severity of suffering depends on the families' ability to adapt and on the specific nature of coping strategies adopted.
- Several coping strategies may have long-term irreversible consequences on transmission of poverty (e.g., deinvesting in children).
- It is critical to identify the most vulnerable households and individuals and apply a rapid and effective response to prevent further deterioration.
- It is critical to understand the most vulnerable households' coping strategies in order to craft both short-term and long-term responses.
- The international community was ill prepared for this crisis in that early-warning, monitoring, and surveillance systems were lacking; not much is known about the impact of the crises on nutrition.
- Global and national responses should support, rather than undermine, the coping strategies adopted by the poor; the response should prevent such households from adopting coping strategies that will cause long-term irreversible damage on human capital formation.
- Targeted nutrition interventions need to be better incorporated into agricultural interventions, social protection programs, income generation projects, and women's empowerment programs; the international nutrition community must work across sectors and integrate nutrition interventions into the global development agenda.

DISCUSSION

This discussion section encompasses the question-and-answer sessions that followed the presentations summarized in this chapter. Workshop participants' questions and comments have been consolidated under general headings.

Remittances

Remittances have become very important in recent times and are particularly affected in the global financial crisis. In many countries that are very dependent on remittances, the WFP has evidence of current remittances decreasing rapidly and drastically for the first time in 20 years. In urban areas, people are either recipients of international remittances, or they themselves send remittances to rural areas. Although the impacts from the financial crisis are primarily seen on middle-income groups, remittance issues are the notable caveat in that those who depend on remittances are usually the most poor and vulnerable.

Eating Out

Counterintuitively, the crisis in West Africa has actually allowed people to keep some diversity in their diet because people are "eating out" more. Typical meals eaten at informal restaurants include some meat, some vegetables, and a larger amount of a staple food. Ironically, eating out seems to be one way people maintain diet diversity as opposed to merely accessing calories from staple foods at home.

Urban Interventions

One workshop participant voiced the concern that interventions in urban areas may actually bring more people to urban areas, seeking that intervention. Different thinking is needed to plan the delivery mechanisms for the urban poor and the rural poor, and the incentives of urban interventions should not be so attractive that they instigate an even more serious problem of urbanization than already exists.

Investment in Rural Agriculture

The belief that the international community should keep food prices artificially low as a response to the current food crisis and ongoing price volatility is controversial. The idea that lower food prices (even if they are kept artificially low) are better than higher food prices makes intuitive sense. Yet this issue needs to be considered from a dynamic perspective. The true negative aspects of the current crisis stem from the lack of investments in rural areas. Agricultural productivity has not increased, and infrastructure development has been ignored. Several workshop participants argued that because food prices were kept artificially low, agricultural investments were neglected. As a result, now a variety of social intervention programs are needed—including transfer programs and poverty relief programs. The underlying issue of a lack of investment in rural agriculture still remains.

REFERENCES

Abebe, D., and N. Singburaudom. 2006. Morphological, cultural and pathogenicity variation of exserohilum turcicum (pass) leonard and suggs isolates in maize (zea mays l.). *Kasetsart Journal—Natural Science* 40(2):341-352.

Action Contre la Faim. 2009. *Feeding hunger and insecurity: The Global Food Price Crisis.* London: Action Against Hunger..

Black, R. E., L. H. Allen, Z. A. Bhutta, L. E. Caulfield, M. de Onis, M. Ezzati, C. Mathers, and J. Rivera. 2008. Maternal and child undernutrition: Global and regional exposures and health consequences. *Lancet* 371(9608):243-260.

Food and Agriculture Organization of the United Nations. 2008. *Cambodia Initiative on Soaring Food Prices.* Cambodia: FAO/WFP.

———. 2009. *More People Than Ever Are Victims of Hunger.* Rome: Food and Agriculture Organization.

Headey, D., and S. Fan. 2008. Anatomy of a crisis: The causes and consequences of surging food prices. *Agricultural Economics* 39(SUPPL. 1):375-391.

International Bank for Reconstruction and Development. 2007. *World Development Report 2008: Agriculture for Development.* Washington, DC: The World Bank.

Jones, G., R. W. Steketee, R. E. Black, Z. A. Bhutta, and S. S. Morris. 2003. How many child deaths can we prevent this year? *Lancet* 362(9377):65-71.

Menzel, P., and F. D'Aluisio. 2005. *Hungry Planet: What the World Eats.* New York: Random House, Inc.

UN. 2004. *World Urbanization Prospects: The 2003 Revision.* New York: United Nations.

Victora, C. G., L. Adair, C. Fall, P. C. Hallal, R. Martorell, L. Richter, and H. S. Sachdev. 2008. Maternal and child undernutrition: Consequences for adult health and human capital. *Lancet* 371(9609):340-357.

von Braun, J., A. Ahmed, K. Asenso-Okyere, S. Fan, A. Gulati, J. Hoddinott, R. Pandya-Lorch, M. W. Rosegrant, M. Ruel, M. Torero, T. van Rheenen, and K. von Grebmer. 2008. *High Food Prices: The What, Who, and How of Proposed Policy Actions.* Washington, DC: International Food Policy Research Institute.

4

Responding to the Crises at the Country Level

Between 2003 and 2008, the world prices of maize and wheat tripled and the price of rice quadrupled (von Braun, 2008). The purpose of this chapter is to describe the range of experiences of various countries in dealing with these dramatic food price spikes as well as with the ongoing economic downturn. Presenters examined the impact of these economic factors on food security and nutrition, and representatives from four countries recounted those countries' responses to the crises. As described by moderator Ruth Oniang'o of the Kenyan Rural Outreach Program, the following presentations helped workshop participants to understand the range of country experiences with the food price and economic crises and their impact on food security and nutrition, as well as country-level responses to these crises.

THE ROLE OF MINISTRIES IN RESPONDING TO THE CRISES AT THE COUNTRY LEVEL

Ruth Oniang'o, Ph.D., Executive Director
Rural Outreach Program, Kenya

People living in urban centers are most immediately affected by economic shocks and are more likely than rural consumers to turn violent in the face of hunger. Such violence tends to influence governments to act quickly because of the likely political implications. The recent food crisis saw outcries in many countries; there were food riots and demonstrations in a number of cities across the globe. These food riots spurred the respective governments into action. Whether these governments acted because they wanted to tackle poverty and malnutrition or because they merely wanted to preserve their own survival is an unanswered question.

Government Interventions

Realizing the implications of rising food prices, many governments instituted food subsidies, imposed price controls, restricted exports, and cut import duties. Some governments increased cash transfers to the hungry, and in some cases there was limited use of feeding and nutrition programs due to their prohibitive costs.

Ministries of health are especially linked with nutrition programs. Many work with the United Nations Children's Fund (UNICEF) to target the vulnerable and severely malnourished, especially children, and provide food rations. Ministries of agriculture are also involved in mitigating the impacts of the global food crisis. These ministries develop long-term strategies to boost agricultural production. In some countries, ministries of agriculture provide subsidies and incentives to farmers and importers of food grains. Ministries of finance have begun temporarily lifting taxes and tariffs on agricultural imports. They can combat the negative effects of the food crisis by redesigning their countries' budgets to meet the new demands of the current emergency.

What Countries Need

Many poor countries cannot manage the effects of food shortages on their own. International organizations need to become involved. Due to safety concerns, such organizations do not engage in places where violence and riots are a reality. When possible, country governments should partner with international organizations to increase targeted programs to the poor and vulnerable; the priority needs to be addressing the emergencies of starvation and destitution. Countries need technical assistance in dealing with these types of emergency situations. The media plays a key role in informing and engaging the global community. In this way, the media can be an ally to country governments when provided with accurate information about the country-specific situation.

Developing countries need to take early warnings seriously. In Kenya, for example, government officials know that crops have failed and that there will likely be no rain for the next 6 months. Yet the government is not acting on this information. The capacity of government officials in developing countries need to be built. Officials need to be empowered to use the information at hand to make life better for their citizens.

At this time, in order to cushion the poor and the vulnerable, technical support is badly needed in dealing with the food price and economic crises. The ministries of health, agriculture, and finance need new skills to deal with emerging problems such as climate change and rising food and fuel prices. It is imperative that government officials begin to recognize and analyze gender dimensions. The answer to Africa's food and governance problems lies with the women of Africa because women produce over 80 percent of the food consumed on the continent, yet receive little in terms of training and resources.

REVIEW OF NATIONAL RESPONSES TO THE FOOD CRISIS

Hafez Ghanem, Ph.D., Assistant Director-General,
Economic and Social Development Department
Food and Agriculture Organization

The crisis of undernourishment in the world is a chronic crisis and a structural problem. Yet in a recent Food and Agriculture Organization of the United Nations (FAO) review of 81 countries' policy responses to the food crisis, all of the policies reviewed were short term in nature. Unless there is a major change in the way governments address hunger both in developing countries and at the global and international level, the problem will remain unresolved.

The Food Security Crisis

The real food price index began increasing in 2002 and rose sharply from 2006 until mid-2008. By mid-2008, real food prices were 64 percent above the levels of 2002, but today prices are still above where they were before all this began. Today, the median price of food in a sample of 55 developing countries is 25 percent higher than a year ago in real terms (Food and Agriculture Organization of the United Nations, 2009a). For many of the poorest people in the world, the food price crisis is still a bleak reality. According to the *OECD-FAO Agricultural Outlook*, jointly published by the Organisation for Economic Co-operation and Development (OECD) and FAO, a decline in real prices is expected (relative to their peak in mid-2008), but over the next decade they are projected to remain above where they were before 2005 by about 10 to 15 percent (OECD and FAO, 2009). There is much uncertainty around this prediction, however, because the future price of oil is unpredictable, and there is now a strong link between oil and food prices (Food and Agriculture Organization of the United Nations, 2008a).

The relationship between oil and food prices is very strong. On the cost side, energy represents a significant share of the cost of agricultural production, particularly in developed countries. On the demand side, with increased biofuel production, the price of oil now helps set a minimum price for food crops because if the price of agricultural commodities used as feedstocks for biofuel production—for example, corn—falls too much relative to oil, it becomes profitable to produce ethanol. If it becomes profitable to produce ethanol, there will be more demand for corn. In this sense, the oil price is now setting a floor price for food products. The future of food prices, then, depends a great deal on what happens to oil prices.

Superimposed on the continuing food price crisis is the financial crisis that brings declining employment, wages, and income. The World Bank and International Monetary Fund data predict a decline in the global economy in 2009: remittances from developing countries are expected to fall 5–8 percent; global trade is predicted to fall by 5–9 percent; a decline of 32 percent in foreign direct investment in developing countries is expected; a decline of 25 percent in foreign assistance is anticipated; and credit will be harder to access (Food and Agriculture Organization of the United Nations, 2009a). Food prices have fallen, but for many people incomes have fallen even further, reducing their access to food.

FAO estimates that more than 1 billion people are hungry in 2009 (Figure 4-1). (These figures consider caloric intake only. They don't take into account the quality or other nutritional content of the food being consumed.) Viewing the estimated numbers of hungry from 1970 through today, one could argue that it is wrong to talk about a food crisis or an undernourishment crisis. The word *crisis* implies that it is temporary, and this graph clearly demonstrates that global hunger is not temporary. It is, instead, an inherent structural problem. While this

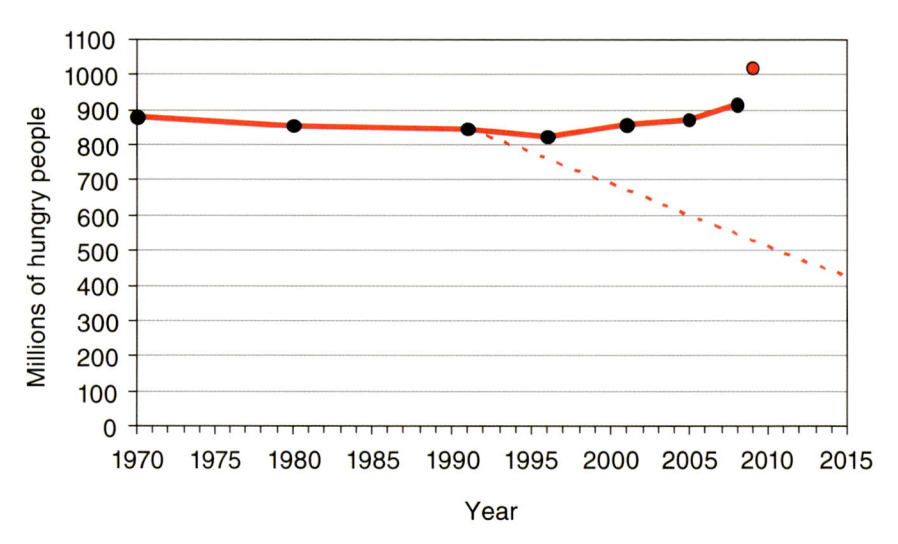

FIGURE 4-1 The number of hungry people in 2009 (1.02 billion).
NOTE: The dotted line indicates the trajectory needed to meet the 1996 World Food Summit goal of reducing the number of hungry people by half by 2015. The FAO measure of food deprivation, referred as the prevalence of undernourishment, is based on a comparison of usual food consumption expressed in terms of dietary energy (kcal) with minimum energy requirement norms. The part of the population with food consumption below the minimum energy requirement is considered underfed (Food and Agriculture Organization of the United Nations, 2008a).
SOURCE: Food and Agriculture Organization of the United Nations, 2009a.

problem has been ongoing since this type of data was first collected, the recent increase in food prices may serve as a call for the international community to realize that something must be done to address the large and growing numbers of hungry people.

International Responses

In July 2009, the G8 made a commitment to mobilize $20 billion over the next 3 years for a comprehensive strategy focusing on sustainable agricultural development. Prior to this announcement, the international response had been mainly a reaction to a particular short-term crisis. Resources were put into safety nets, feeding hungry people and helping farmers with agricultural inputs. While these short-term measures are important, it is also essential to invest in long-term rural development. The $20 billion committed by the G8 indicates a positive shift in policy toward looking at the long-term issues.

The crisis has mobilized others beyond the G8 as well, including other donors and those in the multilateral system. The European Commission gave €1 billion to support inputs of seeds and fertilizers for farmers in developing countries. The initiative grew from the idea that European farmers require fewer subsidies because prices are high; instead, the European Commission chose to subsidize farmers in developing countries. The World Bank also shifted a great deal of resources toward agricultural support, and the World Food Programme's intervention of food aid was essential as an immediate safety net for the most vulnerable in times of emergency.

Developing Country Responses

The specific policy interventions adopted by developing countries in response to the food price crisis are grouped into those that are trade oriented, those that are producer oriented, and those that focus on consumers. Of the 81 countries that FAO reviewed, all took *some* policy measure in response to the increase in food prices.

Trade-Related Measures

Of the 81 countries reviewed, more than 25 instituted export restrictions (a short-term response). When prices of food crops are increasing, a government should encourage its farmers to produce more food crops. Export restrictions, by contrast, keep domestic prices artificially low and reduce farmers' incentives to produce food. Further, export restrictions are not long-term solutions; people find ways of bypassing them. Yet export restrictions are legal under the World Trade Organization's international agreements.

Almost all of the countries in the FAO sample that had tariffs or taxes on imported food products removed them in response to the food crisis. Like export restrictions, these measures further reduce domestic producer prices, but have very little impact on consumers because typically the tax or tariff rate on food is proportionately quite low. For example, if the price has gone up by 60 percent, the removal of a 5 percent tax would not affect consumers much.

Import measures did reduce domestic prices and stabilize domestic markets, but the measures have tended to reduce domestic producer prices (although to a smaller degree than export measures) and deplete the foreign exchange of poor countries.

Production-Promoting Policies

A few countries employed programs to support farmers. While it is important to help farmers increase productivity, a renewed focus by some countries on achieving domestic food self-sufficiency because of volatility in international markets can be costly. For many countries it may be more cost-effective in the long term to purchase food with earnings from other types of exports. Large portions of the world, especially in Asia, the Middle East, and North Africa, will always depend on international trade to ensure food security.

There is an increased (and controversial) interest by cash-rich countries in acquiring land abroad for securing food and fodder. This type of "foreign direct investment in agriculture" can be a very positive, win-win situation if done properly—by partnering with farmers. But if not done in a transparent and participatory manner, it can have adverse impacts on the food security of developing countries.

Other production-promoting measures such as renewed interest in input subsidies, output price support, and grain reserves have been put in place, but many poor countries lack the necessary resources to follow through with these programs.

Consumer-Oriented Policies

Many countries attempted to support the consumers with new policies, either by providing subsidies to consumers or trying to fix prices. Producers were often adversely affected by these consumer-oriented measures. The subsidies, in many cases, turned out to be quite unsustainable; the prices were so high that the cost of subsidizing consumers was also extremely high. Fixing prices provided a disincentive for increased production. Targeted transfer programs to vulnerable groups (food stamps, vouchers, and school feeding) were seen to be more efficient than tax reductions and price subsidies.

Conclusion

Even where they benefit consumers in the short term, the above policy responses do not address the main issues of poverty and food insecurity for the billion people who are undernourished. The last time the world faced a similar situation in such large numbers of undernourished people was in 1974 (when food prices were even higher than in 2008). The world's response at the time was a huge increase in investment in agriculture—the Green Revolution—including funds from developing countries themselves, especially India and Bangladesh, who put a large portion of their own budgets into agriculture and food security. For decades following the Green Revolution, the number of undernourished people in the world was declining. Today the numbers are increasing; the current food crisis is the result of years of neglect.

The policy agenda that lies ahead is an important one. Countries need to design more supportive, sustainable, and long-term food and agriculture policies with expanded investment in food production and research. Governments need to partner with the international community to ensure access to targeted safety nets for vulnerable groups. Additionally, global and national governance around food and nutrition must be enhanced.

The recommendations to invest in various programs or policy measures are primarily technocratic solutions. But it is politicians who make decisions. There is a need to empower the institutions and the governance structure that gives voice to farmers, poor people in rural areas, and the hungry in general, wherever they are.

THE CASE OF MEXICO

Graciela Teruel Belismelis, Ph.D., Professor, Department of Economics
Iberoamericana University, Mexico

Over the past 20 years, demographic changes have occurred in Mexico. Households have become smaller as people are having fewer children, and there are fewer young children. Total household expenditure has been increasing over time; however, there was a decline in household expenditures in 2008 and 2009. Over time, people have become wealthier, with the exception of the financial crisis that occurred in Mexico in 1996. This presentation focuses on food prices, expenditures, consumption, and nutrition in Mexico.

Food Prices and Expenditures

In the past 20 years, the National Food Price Index shows that the price of eggs has increased the most of all foods, marked by a rapid price increase in the past 2 years. In addition, increases have been seen in fats and oils, maize, corn, legumes, fruits, cereals, and tubers. Sugar and sweeteners have had the lowest increase relative to the food price index; as a result, the relative price of soda and alcoholic beverages has actually decreased over time.

At the macro level, Mexican food prices can be compared to international prices. For cereals and grains, the trends are very similar. For fruit and vegetables, prices are slightly higher at the international level. However, the prices of meat, eggs, and milk are much higher internationally. The Mexican government's response to the food price increases has been focused on supporting food producers and transferring cash to families through *Oportunidades* and other programs. The government has been setting prices and quotas, giving credit, subsidizing input prices, and distributing food across the nation. *Oportunidades* is a program that transfers cash to poor families for food—one out of every five families in Mexico is aided by this program.

Mexicans spend approximately 70 percent of their money on nonfood items. The remaining 30 percent is spent on food. Since 1996, the amount of food consumed inside the home has been decreasing while food consumed outside the home has been increasing. The average Mexican diet consists of meats, maize and corn, cereals, nonalcoholic beverages, vegetables, and milk. The primary expenditure is on meats with a much smaller share on fruits and vegetables. This has been consistent over time and has not changed as a result of the crisis. Since 1992, people have been buying more prepared foods. They are also buying more nonalcoholic beverages, primarily sweetened soda, and more dairy products. As the relative prices of soda are decreasing, its consumption is increasing. Negative trends for food expenditures have been seen as well. Fewer vegetables, eggs, and legumes are being consumed.

Impact on Nutrition

Overall, Mexico is winning the fight against stunting. For children under 5 years, the prevalence of stunting has decreased from 1988 to 2006. In 2006, only 12.7 percent of children under 5 years were stunted, as opposed to 22.8 percent only 10 years ago. There is, however, disparity in stunting among different regions of Mexico. The south has the highest prevalence of child stunting, while the north has the lowest. Prevalence in the north has not changed over time, likely because of lower initial levels.

With respect to geographic regions, rural areas have higher stunting rates, although these are declining over time. Decreased child stunting has also been seen across the deciles of distribution. The highest reductions have been in the first deciles, or the poorest of the poor. The reduction in prevalence of stunting in indigenous people in Mexico has occurred at a much lower level; from 1988 to 2006, the rate of reduction in indigenous populations has been about 50 percent lower than the rest of the population (despite the fact that Mexico had special programs targeting indigenous populations).

At the national level, the prevalence of anemia in children under 5 years has decreased from 28 percent in 1999 to 23.7 percent in 2006. There is not a significant difference between anemia prevalence in urban and rural populations. In the past 20 years, the number of obese and overweight adult women in Mexico has increased dramatically and is becoming an epidemic in Mexico. For example, almost 70 percent—7 out of 10 adult women ages 20–49 years in Mexico—are either overweight or obese.

THE GLOBAL FOOD PRICE CRISIS AND FOOD DEVELOPMENT STRATEGY IN CHINA

Fangquan Mei, M.S., Ph.D., Standing Vice President
State Food and Nutrition Consultant Committee, State Council of China

This presentation addresses the impact of the global food price crisis on China's current food situation and offers policy recommendations for grain and food development in China.

China's Grain Situation

China is faced with continued population growth, severe constraints on agricultural resources, a speeding up of market transition processes, and a rapid increase in correlation with the international market. As a result, a series of new problems have emerged regarding China's grain. From 1996 to 1998, the output of China's grain rose to 512 million tons. At that time, a series of policy measures were put in place to stabilize grain production. In 2008, grain production was 519 million tons.

Currently, China's grain reserve is at 150–200 million tons, accounting for 30–40 percent of China's annual grain consumption. The level of the World Food Security Reserve is 17 percent; therefore, China's grain stocks are double the world average level. During the recent food price crisis, domestic grain reserves were sufficient to maintain domestic food security. However, international grain prices and domestic price inflation could not have been avoided, and grain prices did increase. A solution for rising grain prices is direct subsidies to low-income people who have inadequate purchasing power. Such solutions will continue to be necessary because grain prices are predicted to rise to a new level and stabilize at that higher price.

By 2050, the world population will reach 9.5 billion (from the current 6.2 billion); global demand for grain in 2050 will be more than double what it is today. A significant increase in demand for grain will likely lead to another sharp rise in global agriculture and grain prices. This prediction is an important cornerstone of China's grain policy.

Next Steps for China's Grain and Food Development Policy

To ensure China's grain and food security, China needs to implement a series of strategic measures and effective policies. The priorities are discussed in the following sections.

Protect Arable Land and Promote Productivity of Farm Lands

Strict implementation of the basic farmland protection system must be applied and the quality of farmland must be promoted. At present, the total area of crop land with low productivity accounts for 64.58 percent of total farmland area. Currently, a soil-fertilizing project is being carried out with an expanded irrigation area in order to promote farmland productivity.

Increase Utilization Efficiency of Fertilizer and Irrigation Water

The level of fertilizer input is relatively high in China, but it is unbalanced in different regions. The utilization efficiency (UE) of fertilizer is far below the world level, and the average UE of nitrogen and phosphor fertilizers is only 30 percent. In the near future, the adjustment of chemical fertilizers will be needed, with a focus on increasing the benefits of the use of this type of fertilizers.

Water resources in China are severely lacking. For many years, the effective irrigation area of farmland has been about 55 percent of total farmland area, and the UE of irrigation water is an estimated 45 percent. Future policies should emphasize development of water-saving agriculture.

Increase the Fund for Agricultural and Rural Development

In recent years, the central government has increased the fund for agriculture and rural development to a great extent. In 2007, it was increased to 431.8 billion Yuan, and in 2008 it increased to 587.7 billion Yuan. The fund for rural infrastructure construction needs to be increased to about 30 percent from its current 15 percent in order to promote basic productivity.

Promote Innovation Capacity of Science and Technology for Grain and Food

In an effort to promote the innovation capacity of science and technology for grain and food production, the following should be considered a priority:

- Establish and improve China's innovation system for agricultural science and technology.
- Improve the system for knowledge dissemination by making efforts to extend key techniques for grain production to farmers.
- Enhance the capability for preventing and reducing disasters.

Strengthen Safety Controls in the Food Production Process

To strengthen safety control in the food production process, the following is recommended:

- Implement the Pollution-free Food Action Plan.
- Speed up the development of "green" food.
- Increase production of agricultural products.
- Expedite the construction of a traceability system.
- Improve monitoring systems for pesticide and chemicals.

Deepen the Reform of Grain Circulation Systems

To deepen the reform of grain marketing systems, the following is recommended:

- Establish the three-market system: the production area market, the terminal market, and the distribution market.
- Foster and develop food wholesaler, agent, producer, and circulation associations to enhance the organization and management of food circulation.
- Enhance the construction of a network system for grain and food market information at home and abroad to ensure the rapid exchange of supply and demand information.
- Build a smooth transportation channel for grain and food to promote the efficiency of grain circulation.
- Speed up the pace of grain purchase and sale marketization.

Gradually Build a Feasible Grain Reserve System

Speeding up the construction of a mature grain reserves system is important. The national and local reserve grain system should be constructed respectively, focusing on the construction of grain reserves in grain sale-orientated areas. Scholars have acknowledged the distinct differences in national grain reserve levels, ranging from 85 to 250 billion kg in China. Each individual country should select its grain reserve standard based on its own national conditions.

Promote the Regulation Capacity for Grain and Food Import and Export

Grain imports are generally controlled at below 5 percent of grain consumption. From 2000 to 2003, the production of grain in China continuously decreased, but maize and rice exports increased. In recent years, the amount of imported soybeans has grown to more than 30 million tons, primarily to be used as raw materials for the oil industry. The capacity for grain imports and exports has clearly increased in China.

Construct and Monitor an Early Warning System for Grain and Food Security

From 1997 to 2000, the Chinese Academy of Agricultural Sciences, in partnership with other institutions, developed an early warning system for grain and food security in China. The Chinese Center for Disease Control and Prevention and other institutions are also studying and developing a monitoring and warning system for food safety. The analysis system for grain and food monitoring and warning should be established and improved as soon as possible, to provide food consumption, production, and trade with timely and effective information support.

Establish the Coordinated Lead Management System for Grain and Food

More than 10 government departments are involved in the management of grain and food in China. Therefore, adapting to the rapidly developing economic environment and market changes is challenging. A coordinated system is needed to expedite the reform of the government management system for grain and food in China.

Conclusion

In summary, to respond to the global grain and food crisis and domestic inflation, efforts must be made to establish support systems for food security, integrate the development of urban and rural areas, increase the development of grain and food, adjust the agricultural structure and increase the income of farmers, improve the Chinese food and nutrition structure, and support the sustainable and steady development of China's economy and society.

FOOD PRICES, CONSUMPTION, AND NUTRITION IN ETHIOPIA: IMPLICATIONS OF RECENT PRICE SHOCKS

Paul Dorosh, B.A., M.A., Ph.D., Senior Research Fellow
International Food Policy Research Institute

This presentation synthesizes several pieces of work. Although it covers a number of serious problems, it highlights a great deal of progress in Ethiopia as well. The presentation addresses geography, food production, consumption patterns, and measures of nutrition. It then turns to movements in food prices in Ethiopia and their determinants, aspects of the early warning system, and safety nets.

Background

The geography of Ethiopia has important implications for agricultural production, economic growth, and nutritional outcomes. Ethiopia's topography varies widely even across short distances (especially in the highlands), making transport difficult in many areas. The country is also landlocked; Ethiopia is dependent on one major port (Djibouti), and port congestion and high transport and marketing costs are factors limiting trade with external markets.

There is also high variation in rainfall across space and time, and severe droughts have been major factors for widespread famines in 1973–1974 and 1984–1985. In most years, rainfall is adequate for one crop in the central highlands, but rainfall is typically very low in the dry eastern part of the country. Most of the population lives where the majority of crop production takes place—in the highlands of central Ethiopia where four major cereal crops are cultivated: teff (an indigenous crop), wheat, maize, and sorghum. Throughout the country, but especially in the highlands, the population is more concentrated along major road networks. Nonetheless, much of Ethiopia is remote. Forty-five percent of the population lives more than 5 hours from a city of 50,000 (Kedir and Schmidt, 2009).

Over the past several years, production of the major cereals has increased substantially because of both area and yield increases, from 10 million tons in 2003–2004 to more than 14 million tons in 2008–2009. Moreover, according to the national household income and consumption surveys, the percentage of people below the poverty line decreased from 45 percent in the mid-1990s to 39 percent in 2005. The drop in rural poverty in this period was especially pronounced—from 48 to 39 percent. In the urban areas (where about 18 percent of the population reside), the poverty rate was slightly higher in 2005 (36 percent) than in 1995 (33 percent), although still lower than the rural poverty rate (Government of Ethiopia, various years).

Cereals account for about 60 percent of total calories consumed in Ethiopia. The poorest 20 percent of people in the country consume only 1,672 calories per person per day on average, with three-quarters of calories derived from cereals and enset ("false banana") (Table 4-1). Moreover, with cereals and enset accounting for such a large share of the daily calories for most household groups, consumption of vegetable and animal protein, fats, and oils is generally very low. Nonetheless, there is a wide variation in the composition of cereal consumption across household groups. For example, teff accounts for 30 percent of total calorie consumption in urban areas, but only 9 percent in rural areas where other cereals and enset account for three-quarters of calorie consumption.

Because of the relatively low total food availability, malnutrition in Ethiopia is extremely high: an estimated 46 percent of the population in Ethiopia consumes less than the minimum calorie requirements (Schmidt and Dorosh, 2009). By comparison, only 32 percent of the population is malnourished by this standard in

TABLE 4-1 Ethiopia: Household Calorie Consumption by Source, 2004–2005 (kcal/person/day)

	Teff	Wheat	Other Cereals	Enset/Root Crops	Pulses/ Oilseeds	Animal Products	Other	Total
National	248	266	832	234	214	76	215	2,086
Urban	588	181	495	64	285	85	240	1,937
Rural	192	280	887	262	203	75	211	2,110
Expenditure Quintiles								
Q1	173	215	708	174	148	61	192	1,672
Q2	224	259	812	208	192	67	202	1,964
Q3	225	275	910	256	213	81	205	2,163
Q4	273	304	907	252	247	74	220	2,277
Q5	349	280	828	282	273	98	258	2,367

SOURCE: Calculated from Government of Ethiopia Central Statistics Agency (CSA) Household Income, Consumption and Expenditure Survey (HICES) 2004/2005 data.

neighboring Kenya. Ethiopia also has a high prevalence of underweight children, although the under-five mortality is relatively low.

Determinants of Food Prices in Ethiopia

In spite of increases in cereal production, real prices of cereals (i.e., nominal prices adjusted for overall inflation in the economy) trended upward from the early 2000s through 2006. Nominal and real prices then surged in 2007 and 2008. This jump in prices was caused in part from poor rains in early 2008 that increased expectations of a poor harvest in late 2008 (which fortunately did not come to pass). Total production in the 2008–2009 main season ultimately was good, and nominal and real prices dropped sharply at the end of 2008.

Although these price rises more or less coincided with cereal price increases in international markets, domestic factors largely account for the price increases. Among the four main cereal staples, wheat is the one commodity in which Ethiopia has significant international trade, and there is a plausible argument that there would be a transfer of international prices to the domestic wheat market. Comparisons of domestic prices with import parity prices (the price of wheat in the international market plus transport and marketing costs to wholesale markets in Ethiopia), however, show that for much of the period between 2000 and 2009, it was not profitable for private traders to import wheat (Dorosh and Ahmed, 2009).

From 2000 to 2005, supply of wheat from domestic production and food aid were sufficient to keep domestic wheat prices below import parity levels (and above export parity level),[1] so there was no incentive for the private sector to trade wheat internationally (Figure 4-2). Then, from 2005 to 2007, domestic demand for wheat exceeded domestic supply (including food aid) at import parity prices, making private-sector imports profitable. During this period, changes in international prices directly affected domestic wheat prices.

However, in 2007–2008, when international prices rose steeply, domestic wheat prices showed only a moderate increase in real terms (and were thus substantially below import parity) largely because another good cereal crop harvest in Ethiopia allowed domestic supply (production plus food aid imports) to meet demand at moderate price levels. During this period, it was no longer profitable to import wheat into Ethiopia. Only in early 2008, as overall inflation in Ethiopia accelerated, did wheat prices rise substantially, surpassing import parity levels. This domestic price rise did not trigger price-stabilizing private-sector imports, however, because foreign exchange restrictions—put into place to minimize the balance of payments deficit—prevented importers from taking advantage of the otherwise profitable trading opportunity.

---

[1] Export parity prices are international prices less transport and marketing prices to wholesale markets.

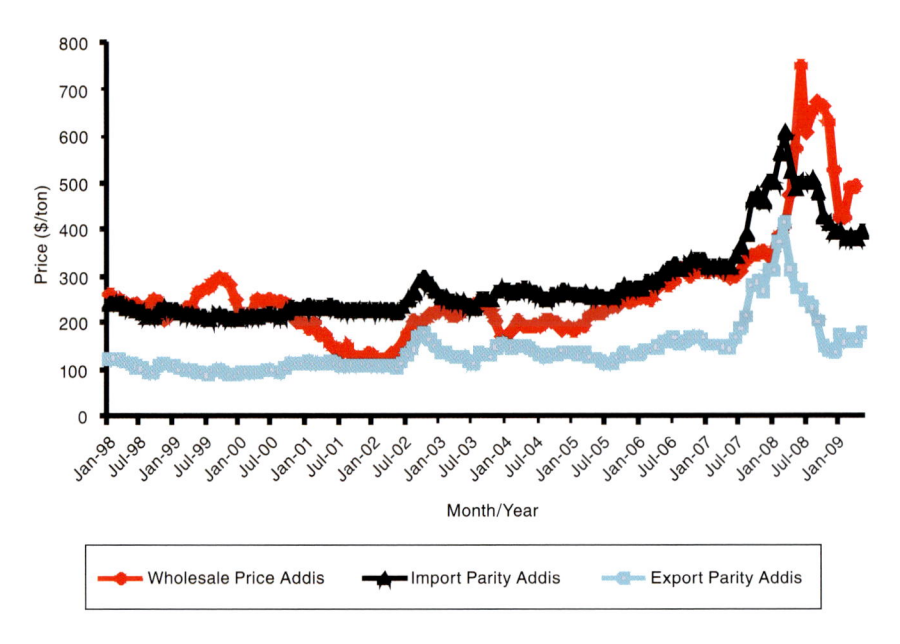

FIGURE 4-2 Domestic and international wheat prices in Addis Ababa, Ethiopia, 1998–2009.
SOURCE: Dorosh and Ahmed, 2009.

Finally, from July to October 2008, government sales of its own wheat imports successfully reduced domestic market prices. Because this wheat was sold in rationed amounts at prices substantially below market price levels, sizeable rents (excess profits) accrued to those with access to wheat imports at official prices.

Early Warning Systems and Livelihoods Analysis

Although cereal price movements in Ethiopia were largely determined by domestic production and policies, such factors as national and regional production, income, and price shocks still strongly affected household nutritional outcomes, depending on the structure of household incomes (their livelihoods) as well as their food consumption patterns. To better understand and anticipate the severity of these shocks on food security, the Livelihoods Integration Unit was incorporated within the Ministry of Agriculture and Rural Development as part of the government's early warning system.

Baseline data on various household groups have been gathered for 173 livelihood zones using key informant questionnaires. These data are used to estimate potential effects of price and income shocks on household access to food. Such an analysis enables a more disaggregated analysis of household food security that can identify problems that may not be captured in regional analyses. For example, this analysis highlighted the serious effects of production losses of enset and sweet potatoes on some households in pockets of southern Ethiopia in 2008.

Safety Nets

Ethiopia introduced the Productive Safety Net Program (PSNP) in 2005 as a way to provide access to food for poor households and at the same time, help build public infrastructure and household assets that could raise household incomes in the medium term. Initially, the PSNP was targeted to cover 264 food insecure *waredas* (subdistricts), primarily in drought-prone areas. A public works scheme was created where it would pay workers in cash or in-kind for their labor on labor-intensive projects designed to build community assets. It also provided direct support to labor-scarce households including those whose primary income earners are elderly or disabled. In addition, a complementary program, the Other Food Security Program, was created to provide at least one productivity-enhancing transfer for service, including access to credit and agricultural extension services.

In 2006, an evaluation of the Productive Safety Net Program showed that effects varied by household participation (Gilligan et al., 2008). The households that received at least half of the amount of transfers that they should have according to the design of the program had a reduced likelihood of having a very low caloric intake and increased their mean caloric availability by 183 calories per person per day. However, many people in the program received only a fraction of the normal transfer. For these households, there was almost no effect of participation in the PSNP. The greatest effect was seen in participants in the Other Food Security Program. For them, the calorie increase was larger (230 calories per person per day), and they were more likely to be more food secure, to borrow for productive purposes, to use improved agricultural technologies, and to operate their own nonfarm businesses. The initial results suggest that programs that not only provide a transfer but also provide assets or some kind of technological assistance can have a significant positive impact on household food security in Ethiopia.

Conclusions

Over the past two decades, Ethiopia has made impressive progress in enhancing food security by increasing domestic production, investing in infrastructure to improve market efficiency, improving early warning systems, and launching the Productive Safety Nets Program. The recent price shocks in Ethiopia can be traced mainly to domestic factors, including rapidly increasing overall incomes and cereal demand, rather than to shocks in the international market. Moreover, in the context of growing per capita incomes and improved cereal market efficiency, these shocks did not represent a major threat to household food security for the vast majority of households in Ethiopia. Nonetheless, many households remain vulnerable to food production shocks caused by droughts or disease that may be specific to small regions within the country. To further enhance food security for households throughout the country, continued high growth in agricultural production and incomes of poor households will be required, along with expansion of successful household-level interventions such as the Other Foods Security Program and targeted nutrition efforts.

BANGLADESH CASE STUDY

Josephine Iziku Ippe, M.Sc., Nutrition Manager
United Nation's Children's Fund, Bangladesh

This presentation explores the effects of the global food price crisis and its impact on nutrition, policy responses, and suggested necessary actions in Bangladesh.

Background

Bangladesh is a country with a population of about 150 million people. The density of the population is around 952 per square kilometer, the highest in the world, except for some city states, including Hong Kong. The rural population comprises about 76 percent of the total population.

Even before the spike in food prices the nutritional situation included persistently elevated levels of underweight, chronic malnutrition (stunting), and acute malnutrition (wasting), as shown in Figure 4-3. Bangladesh has nearly 9 million stunted children, and it ranks fourth, after India, Indonesia, and Nigeria, out of 36 countries with stunting prevalence greater than 20 percent. These countries total 90 percent of the estimated number of 178 million globally stunted children (Black et al., 2008).

In the 1990s, rates of undernutrition reduced progressively; however, the trend was not enough to put Bangladesh in reach of Millennium Development Goal 1, with the target of reducing underweight from 66 percent to 33 percent by 2015. The last nationally representative nutrition survey in Bangladesh occurred before the increase in food prices and was conducted over the period that coincides with the monsoon season, a time that corresponds to the lean period in the country (BBS/UNICEF, 2007). The survey found that 41 percent of the children under 5 years were underweight, and 43.2 percent suffered from stunting.

Infant and young child feeding practices in Bangladesh are also matters of concern. According to the Bangladesh Demographic and Health Survey (BDHS) 2007, the exclusive breast-feeding of children under 6 months has not improved

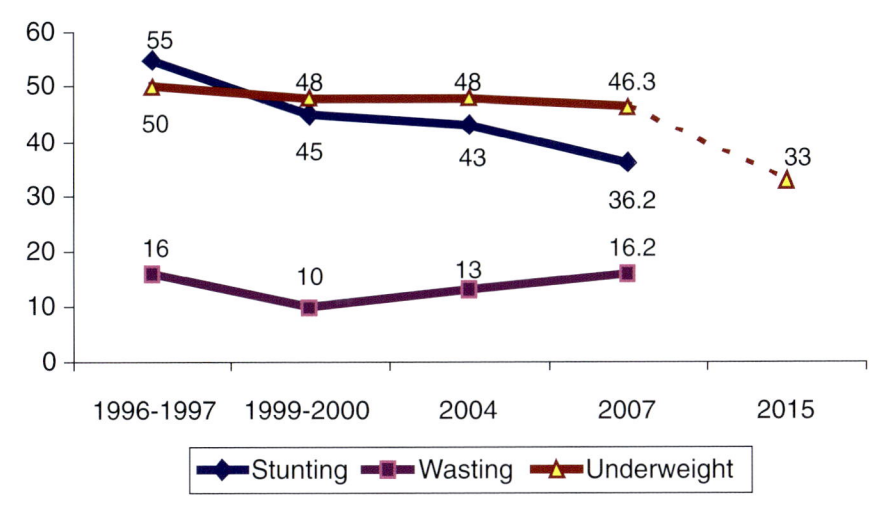

FIGURE 4-3 Trends in nutritional status of children age 6–59 months, 1996–2007, Bangladesh.
SOURCE: USAID, 2009.

in the past 15 years; the figure has remained static at 42–45 percent since 1993–1994. The nutritional status of women as measured by body mass indices showed that 30 percent of mothers were chronically malnourished. Although this finding was an improvement from 34 percent, the figure still remains high and implies a high risk for poor nutritional status in their children (USAID, 2009). Nutrition surveys carried out during the past decade all confirm that women in Bangladesh have low-quality diets. Micronutrient deficiencies among children and women in Bangladesh are major public health problems. Findings from the Helen Keller International (HKI)/Institute of Public Health Nutrition (IPHN) national anemia survey in 2004 showed that 68 percent of children aged 6 to 59 months were anemic with the highest prevalences among infants aged 6 to 11 months (92 percent). In this same study, the prevalence of anemia was lower in children who had been dewormed (BBS/UNICEF, 2004). Anemia has been found in 46 percent of pregnant women and 39 percent of nonpregnant women (HKI, 2006) and in more than one-third of adolescent girls (39.7 percent), predominantly a result of depleted iron-stores during pregnancy and lactation, the consequence of repeated infections, and poor intakes of food rich in iron and folic acid.

The government has developed various policies, strategies, and organizational structures to address malnutrition among women and children, but the delivery of nutrition services remains weak. Nutrition programming is hampered by a lack of coordination among the many actors involved, limited institutional capacity, and, in most of the country, inadequate linkages between the government's health care structure and communities. There is presently no national body with full responsibility and authority for coordinating nutrition activities, and there is no overarching framework existing for the many different types of activities that are underway.

An important dynamic in Bangladesh that undermines nutritional outcomes is seasonality. Levels of malnutrition (acute and underweight) follow a seasonal tendency, increasing during the summer months and decreasing in the winter months, reflecting increases in morbidity and restricted access to food resources in summer months. Diarrhea and acute respiratory infections are major causes of illness, especially in children. Diarrheal disease has been repeatedly linked to increased risk of malnutrition, underpinned by conditions such as lack of clean water, poor sanitation, and inadequate health services. Therefore, a national nutrition policy and integration of nutrition programs is required with more attention to nonfood-based strategies and using a multisectoral approach to coordinate activities.

Global Food Price Crisis

From 1995 to 2009, there was an increase in food prices; however, there was a sharp peak in January 2007 (Figure 4-4). After this the price dropped. Although the price has stabilized, it is still almost 23 percent higher than it was in 2006. Issues affecting prices at the regional level include trade barriers, especially with India, and export bans. The large flood of 2007 also affected food prices. Despite this, in 2007, the percentage of food grain imports dramatically increased and

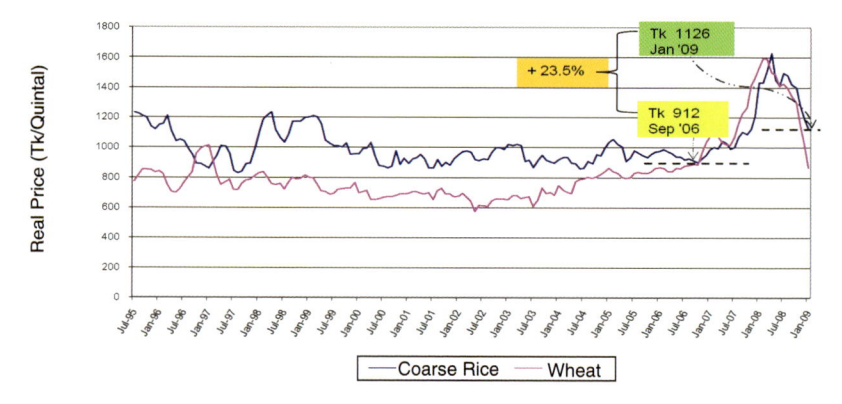

FIGURE 4-4 Household Food Security and Nutrition Assessment in Bangladesh: November 2008–January 2009.
SOURCE: The United Nations Children's Fund (UNICEF)/World Food Programme (WFP)/Institute of Public Health Nutrition (IPHN), 2009.

reached 6 percent of total imports compared to 3 percent in previous years. The food price shock clearly worsened the food security situation in 2008 with 40 percent of households in Bangladesh reporting that they were greatly affected.

Due to the higher food prices, a majority of households in Bangladesh lost purchasing power. In 2008, the real monthly income per household decreased by 12 percent when compared to 2005 incomes. Real wage rates remained stable while the terms of trade (daily wage/rice price) further decreased in 2008. Moreover, expenditures (particularly for food) increased to an unprecedented level of 62 percent of the total expenditures for households. Overall, about one in four households nationwide was affected. These households are defined as being food insecure based on food consumption scores. Those livelihood groups that were most affected include nonagricultural laborers, agricultural laborers, and casual laborers. Garment factory workers are included in the nonagricultural laborers group, and because Bangladesh depends very heavily on garment exports, that livelihood group was greatly affected.

Impact on Nutrition

The joint World Food Programme (WFP), UNICEF, and IPHN Household Food Security and Nutrition Assessment (UNICEF, WFP, and IPHN, 2009) conducted from November 2008 to January 2009 to establish the impact of high food prices found that the overall acute malnutrition (13.5 percent) and underweight (37.4 percent) indicators remained high despite the fact that this survey was undertaken during the harvest season (*Household Food Security and Nutrition Assessment*, in press). Even within this "stable" context, conservative caseload estimates of acutely malnourished children are projected as approximately 2.2 million children, of which more than 0.5 million of these children are severely acutely malnourished and at increased risk for mortality (Black et al., 2008). The Household Food Security and Nutrition Assessment 2009 also found a slight increase in chronic malnutrition or stunting (48.6 percent) when compared to the BDHS 2007 (46 percent).

The food price spikes in Bangladesh have meant that vulnerable children and women are not being provided with the essential dietary requirements and micronutrients necessary to prevent detrimental effects on their nutrition status. The households' expenditures on food purchases are insufficient to provide the quality diets necessary to meet optimum requirements for the growth, development, and nutritional well-being of these children and women who are often already nutritionally compromised. In this way, the price increases will have long-term impacts (Sanago, 2009; Save the Children UK, 2009). The assessment found that wasted, underweight, and stunted children were more likely to have a malnourished mother, demonstrating the importance having a healthy mother has toward decreasing the risk of undernutrition in children. Children of acutely malnourished mothers were 1.8 times more likely to suffer from acute malnutrition, 1.3 times more likely to be stunted, and 1.7 times more likely to be underweight.

Policy Responses

Bangladesh has an extensive social safety net with multiple programs and objectives. Most programs are administered by the government of Bangladesh, but nongovernmental organizations (NGOs) and other nongovernment bodies also play significant roles as service providers. Primarily, assistance is in the form of food or cash-based transfers and targeted at poor and vulnerable groups.

The largest social safety net programs typically operate in rural areas and are generally food based and government administered. Most are linked to the government's Public Food Distribution System. Although the Public Food Distribution System has numerous programs and channels through which food assistance is provided, the majority of assistance (i.e., approximately two-thirds of the total food distributed during fiscal year 2007–2008) is provided through such efforts as Vulnerable Group Development, Vulnerable Group Feeding, Gratuitous Relief, Test Relief, Food-for-Work, and Open Market Sales (Food and Agriculture Organization of the United Nations, 2008a).

Through the local consultative group, FAO also coordinated a policy discussion with the Bangladesh government that focused on a number of food security, agriculture, and rural development issues, while the government procured additional food stocks and subsidies for farmers, especially fertilizer. Overall, the interventions made in response to the increase in food prices were focused on policy and safety net programs, subsidized food distribution, and limited-scale food aid. There were no interventions that specifically focused on improving the nutritional status of vulnerable groups.

Next Steps

The following actions can help to address the negative impacts of the global food crisis in Bangladesh:

- Enhance the efficiency and effectiveness of the social safety net system; expand coverage in areas of high malnutrition and food insecurity and emphasize better targeting.
- Provide cash interventions when food is abundant, accounting for seasonality and market availability; otherwise, targeted food assistance should be provided.
- Support investment in food marketing and storage infrastructure (e.g., warehouses for larger stocks).
- Promote open trade policies within the region, avoid policies that result in trade barriers, and expand and accelerate social protection programs.

- Invest more and build upon existing information systems for monitoring and surveillance and for early warning for early actions.
- Address the large numbers of acutely malnourished children by managing acute malnutrition at both the facility and community levels.
- Develop local production capacity for ready-to-use food; reduce costs and increase the possibility of future government-allocated resources for sustaining the programs.
- Improve optimal infant and young child feeding, emphasizing maternal and community participation.
- Emphasize micronutrient-enriched foods and diet diversity in food assistance interventions, food security, and nutritional programs.
- Strengthen health and hygiene promotion to prevent and treat diarrheal disease, respiratory infections, and fever.
- Harmonize, develop, and standardize national survey guidelines to enable data quality and comparability.

DISCUSSION

This discussion section encompasses the question-and-answer sessions that followed the presentations summarized in this chapter. Workshop participants' questions and comments have been consolidated under general headings.

Case Study Analyses

The four country case studies (Mexico, China, Ethiopia, and Bangladesh) presented during the workshop were not comprehensive analyses of any of the countries. The presenters used a framework to look at food price data. Most presenters learned that, for the time being, it is too early to glean major nutritional impacts from the data. Typically, the most recent data are from 4 or 5 years ago, or at best 2 years ago, but the acute crisis occurred over the past 1 to 2 years. Before any associations are drawn about nutritional impacts of the food price and economic crises, more data are needed from these particular countries.

No Need to Wait for More Data

Several speakers and workshop attendees agreed that while the international nutrition community welcomes more data and improved surveillance to determine the nutritional impacts of the current and ongoing food price and economic crises, increased focus could simultaneously be put toward mitigation of hunger and undernutrition. The problems are known, and the *Lancet* series describes effective strategies for mitigation that should be implemented at all levels—local, national, and global (Bhutta et al., 2008).

Some workshop participants argued that the international nutrition community should focus *more* on mitigation than on analysis. They felt that analysis is important, but it should not get in the way of looking at the best strategies to solve problems, including long-term and short-term solutions. One attendee went so far as to say, "Data are no good if you are not going to use it." Another agreed that continually analyzing, collecting, analyzing, and collecting feels a bit like reinventing the wheel, especially since similar analyses have been done in the past that are still useful today as this is a chronic crisis. It is very important to learn from what has worked and what has not worked in the past.

The Role of Different Ministers at the Country Level

How can the international nutrition community get the health, agriculture, and finance communities and ministers to talk to each other? At the international level, each works in isolation in responding to nutrition. For example, if FAO goes to a country, it talks to a minister of agriculture while the World Health Organization talks to a minister of health. What can be done to facilitate all the different ministers to work together?

To accomplish this goal, it was suggested that political will needs to be mobilized at two levels: the country level and the international level. At the country level, governments have neglected the issue of nutrition for years. The people most affected by food scarcity and undernutrition do not have much voice in the political process, so national governments are not pressured to act.

At the global level, too, coordination and a global political will are needed. This is challenging because controversial political issues between governments of both the "north" and the "south" arise, such as Organisation for Economic Co-operation and Development subsidies and their effect on developing countries; issues of biofuel policies and their effect on food security and development; export restrictions; issues of public investment and aid to food-insecure areas; codes of conduct for private investment; and land-grab issues.

These are important and difficult questions. However, the call by the G8 for a global partnership on food and agriculture is building momentum. Effective coordination among ministries is a challenge in many countries, but with the above focus on both national and international levels, such coordination may be attainable.

REFERENCES

BBS/UNICEF. 2004. *Anaemia Prevalence Survey of Urban Bangladesh and Rural Chittagong Hill Tracts 2003.* Dhaka: Bangladesh Bureau of Statistics and United Nations Children's Fund.
———. 2007. *Child and Mother Nutrition Survey 2005.* Dhaka: Bangladesh Bureau of Statistics and United Nations Children's Fund.
Bhutta, Z. A., T. Ahmed, R. E. Black, S. Cousens, K. Dewey, E. Giugliani, B. A. Haider, B. Kirkwood, S. S. Morris, H. Sachdev, and M. Shekar. 2008. What works? Interventions for maternal and child undernutrition and survival. *Lancet* 371(9610):417-440.

Black, R. E., L. H. Allen, Z. A. Bhutta, L. E. Caulfield, M. de Onis, M. Ezzati, C. Mathers, and J. Rivera. 2008. Maternal and child undernutrition: Global and regional exposures and health consequences. *Lancet* 371(9608):243-260.

Dorosh, P., and H. Ahmed. 2009. *Foreign Exchange Rationing, Wheat Markets and Food Security in Ethiopia.* Addis Ababa: International Food Policy Research Institute (IFPRI).

FAO and WFP. 2008. *FAO/WFP Crop and Food Supply Assessment Mission to Bangladesh.* Rome: Food and Agriculture Organization and World Food Programme.

Food and Agriculture Organization of the United Nations. 2008a. *FAO Methodology for the Measurement of Food Deprivation.* Rome: Food and Agriculture Organization.

———. 2008b. *The State of Food and Agriculture.* Rome: Food and Agriculture Organization.

———. 2009a. *Crop Prospects and Food Situation.* Rome: Food and Agriculture Organization.

———. 2009b. *More People Than Ever Are Victims of Hunger.* Rome: Food and Agriculture Organization.

Gilligan, D. O., J. Hoddinott, and A. S. Taffesse. 2008. *The Impact of Ethiopia's Productive Safety Net Programme and its Linkages.* Washington, DC: International Food Policy Research Institute.

Government of Ethiopia. Various years. *Household Income Consumption and Expenditure Survey (HICES).*

HKI. 2006. *The Burden of Anemia in Rural Bangladesh: The Need for Urgent Action.* Dhaka: Helen Keller International.

Household Food Security and Nutrition Assessment. In press (unpublished). WFP, UNICEF, and IPHN.

Kedir, M., and E. Schmidt. 2009. *Urbanization and Spatial Connectivity in Ethiopia: Urban Growth Analysis using GIS.* Addis Ababa: International Food Policy Research Institute (IFPRI).

OECD and FAO. 2009. *OECD-FAO Agricultural Outlook 2009-2018.*

Sanago, I. 2009. *Rapid Assessment of the Impact of the Global Financial Crisis in Bangladesh.* Rome: World Food Programme.

Save the Children UK. 2009. *How the Global Food Crisis Is Hurting Children: The Impact of the Food Price Hike on a Rural Community in Northern Bangladesh.* London: Save the Children United Kingdom.

Schmidt, E., and P. Dorosh. 2009. *A Sub-National Food Security Index for Ethiopia.* Addis Ababa: International Food Policy Research Institute (IFPRI).

UNICEF, WFP, and IPHN. 2009. *Household Food Security and Nutrition Assessment (HFSNA).* Dhaka: World Food Programme, United Nations Children's Fund, and Institute of Public Health Nutrition.

USAID. 2009. *Bangladesh Demographic and Health Survey 2004.* Dhaka and Calverton, MD: National Institute of Population Research and Training, Mitra and Associates, and ORC Macro International.

von Braun, J. 2008. *Food and Financial Crises: Implications for Agriculture and the Poor.* Washington, DC: International Food Policy Research Institute.

5

A Role for Nutrition Surveillance in Addressing the Global Food Crisis

The food price and economic crises have highlighted the need for collecting data in order to understand the effects of these phenomena on populations and make decisions to improve the situation. There are a variety of tactical measures and approaches to nutrition surveillance that will be explored in this chapter. Workshop presentations discussed an array of nutrition surveillance systems and lessons learned. This chapter considers what roles nutrition surveillance might be able to play in the future, including an investigation of the separate capacities of various agencies and specific projects. A number of presenters spoke of the need to aggregate data, compile it quickly using new technologies, and deliver it to the food security and nutrition community for decision making at the program and policy level. As described by moderator Keith West of Johns Hopkins Bloomberg School of Public Health, the following presentations helped to encourage a broad discussion of nutrition surveillance, including existing nutrition surveillance systems, their capacity to monitor food price fluctuations, and the gaps and needs for improved surveillance.

NUTRITION SURVEILLANCE IN RELATION TO THE FOOD PRICE AND ECONOMIC CRISES

John Mason, Ph.D.,[1] Professor, Department of International Health and Development
School of Public Health and Tropical Medicine, Tulane University

Nutrition Surveillance: Making Decisions to Improve and Protect Nutrition

Nutrition surveillance means to watch over and make decisions that will lead to improvements in the nutrition of populations (FAO/WHO/UNICEF, 1976). Nutrition information—when appropriately linked to interventions, policies, and programs—can help mitigate malnutrition, particularly in developing countries (Figure 5-1). The current economic and food crises serve as stimuli for the world to be concerned about nutrition and perhaps can foster movement in the direction of positive change for nutrition surveillance.

In public health, the "surveillance cycle" is well established and understood. One starts at the point of collecting data, the analysis leads to decisions on action, the action is implemented, the situation is followed, and new data can be used to monitor and evaluate that situation at the same time as detecting new problems (Figure 3-1). This surveillance process applies in the area of nutrition, usually at the population level but may also refer to individuals or households.

Do We Really Know How Many People Are Hungry? (And What Does That Mean?)

The official Food and Agriculture Organization of the United Nations (FAO) estimates from 2009 show that the number of the world's "hungry" has hit 1 billion (Food and Agriculture Organization of the United Nations, 2009). This estimated increase in the world's hungry is similar to that seen around the year 1970, at the time of another world food crisis. Yet hunger is a very "lazy" indicator; its meaning is not very clear and not often explained, and its method of calculation is suitable only to inform that the hunger situation is getting worse.

The true definition of this "hungry" indicator is the number of people who during the course of the past year did not on average get enough food energy to maintain moderate activity and body weight (UN ACC-SCN, 1993). The calculation of this indicator takes the estimated calorie availability at a national level (dietary energy supply [DES]) and sets that together with an estimate of the coefficient of variation (CV) of the distribution of consumption and estimates the number below a certain cut point. It does not take into account a number of

---

[1] The following people contributed to Dr. Mason's presentation: Megan Deitchler, M.P.H., FANTA; Marito Garcia, Ph.D., The World Bank; A. Sunil Rajkumar, Ph.D., The World Bank; and Roger Pearson, M.A., UNICEF.

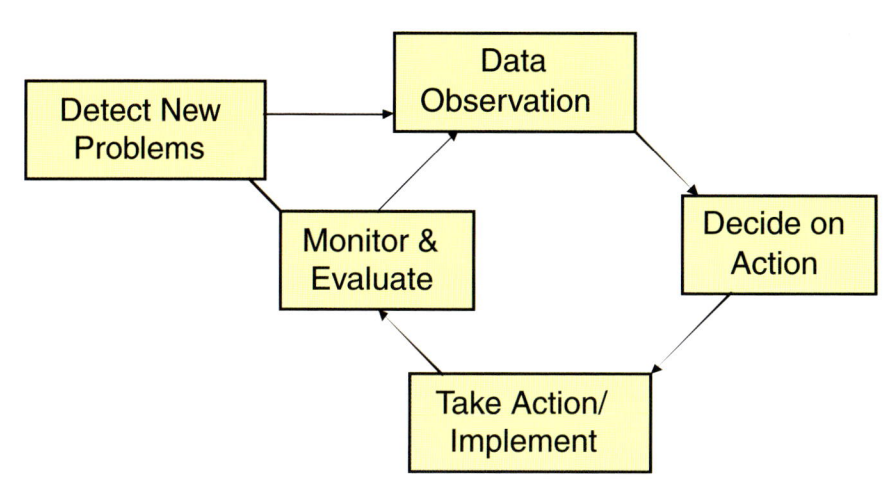

FIGURE 5-1 Surveillance cycle.

factors, such as that most people have a very high degree of correlation between intake and requirement: most individuals do not change weight very much over the course of a year. The international nutrition community should consider whether this hunger indicator is appropriate as the primary means for assessing global food deprivation and undernutrition. A new set of indicators or measurements could be adopted to keep track of the global situation regarding undernutrition, hunger, and malnutrition.

The ability to understand the number and percentage of households that experience hunger would be extremely useful. The methodology needs to evolve towards adding other measures to the existing FAO DES/CV approach to assessing the problem. Suitable methods were previously assessed (Food and Agriculture Organization of the United Nations, 2002) (Table 5-1). One promising method (the fifth listed) involves qualitative data collection—three simple questions about the experience of hunger in households, referred to as the "household hunger scale"—which has now been tested in several countries and could be applied more widely (Deitchler et al., 2009). Other existing measures of the dimensions of hunger are more complicated and expensive to apply. The international community has continued to rely on the first of these indicators; indeed, it can be argued that the current hunger indicator may actually be *obstructive* to progress in hunger assessment because other methods have not been developed as fast as they would have been, were there less reliance on this first (DES/CV) indicator.

TABLE 5-1 Suitability of Different Measures for Trend Estimation

Method	Suitability of Trend Analysis	Dimensions of Hunger Measured
FAO DES/CV	Only reflects DES change as CV held constant; only method available for all countries and all years	Energy intake, with averaged adequacy at population level
Household income and expenditure survey	Potentially suitable when there are repeated comparable large-scale surveys	Energy intake, with some household adequacy; economic aspects (e.g., employment, wages, food prices) also possible
Food consumption and individual intake	Repeated comparable large-scale surveys very rare and expensive	Energy, better chance of relating to requirement, hence adequacy
Anthropometry	Suitable and widely used for trends, but does not measure only (or even) food security	Some aspects of health; changes often related to food access changes
Qualitative methods	Probably very suitable within country, but cross-country comparisons need more work	Suffering, behavior, and economic activity may be assessed

SOURCE: Food and Agriculture Organization of the United Nations, 2002.

Surveillance Inextricably Linked to Interventions

Although data from surveillance systems may be interesting in its own right, clearly the ultimate goal (and inextricable partner) of surveillance data is the link to, and triggering of, appropriate interventions and successful implementation of programs. The interventions to protect nutrition, referred to in a recent article by the UN's Standing Committee on Nutrition (SCN), range from safety nets to a series of essential nutrition interventions, usually best implemented through community-based programs and primary health care. A number of well-established and evolving safety net interventions can mitigate the negative effects of the current food situation, particularly:

- Conditional cash transfers,
- Unconditional cash transfers,
- Food and nutrition programs (school feeding, micronutrient and food supplement distribution),
- Price subsidies for food or energy,
- Public works employment (cash or food), and
- Fee waivers.

These programs, particularly the conditional cash transfers (CCTs), are widespread and are increasing rapidly. One estimate shows 137 CCT programs in 37 countries in sub-Saharan Africa (*Social Safety Nets: Moving Forward for an African Agenda*, 2009). CCTs have a dual or multipurpose rationale: one is to provide income support to needy people; the second is to encourage behaviors that could improve human capital, nutrition, and the education of children. In this sense, the conditionality is to ensure that children are immunized, go to school, or are cared for in ways regarded as beneficial. When public expenditures, infrastructure, and public services fail to reach the very poor, CCTs offer a method of targeting the poor and vulnerable. For these reasons, CCTs may be an important way to foster the development of human capital, as well as healthier, better developed, and more properly nourished children.

How in Theory Would Better Nutrition Surveillance Work?

What are the problems that improved nutrition surveillance could solve? Interventions are inadequate and untimely in preventing worsening malnutrition caused by rising food prices, increasing unemployment, and reduced public or private services. These crisis situations have a particular effect on certain vulnerable groups including pregnant women (with irreversible effects in utero to the unborn child) and the urban poor. The international community, then, needs to think about interventions that can take place quickly and immediately, at least for certain highly vulnerable groups. Then, resources must be allocated to organize and strengthen effective programs and to evaluate whether these programs are having the intended effect.

Surveillance and data collection usually focus on the population level; increasingly, however, surveillance may apply to the individual level as well. There are three types of *population*-level data that are very pertinent to the present economic situation:

1. The real price of food—the ratio of the food price index changes to the general price index—does predict quite well when there is difficulty getting food (when there is food insecurity) and is likely correlated with malnutrition.
2. The "household hunger scale" is a second method of assessing food insecurity or hunger.
3. The third method is tracking the prevalence of malnutrition, usually assessed by anthropometry (the study of human body measurements, especially on a comparative basis). Malnutrition is the result of a number of different factors. It is not the same thing as food insecurity, although it tends to track in the same direction.

The "household hunger scale" could be included in many household surveys (e.g., Demographic and Health Surveys). It uses a short set of questions:

- In the past four weeks, was there ever no food to eat of any kind in your household because of lack of resources to get food?
- In the past four weeks, did you or any household member go to sleep at night hungry because there was not enough food?
- In the past four weeks, did you or any household member go a whole day and night without eating anything because there was not enough food?

The results of such questions are sensible and easy to interpret; they appear relevant and easily communicated across different cultures (Deitchler et al., 2009).

At the *individual* level, eligibility criteria for cash transfers and the numbers of eligible recipients give an indicator as to levels of food insecurity, malnutrition, and hunger. Data from screening children can give a real-time estimate of the number of children who are malnourished. There are other indicators, such as the number of people who participate in public works (paid in cash or food), which can serve as gauges of hunger. The eligibility for access to programs could provide a new set of information and needs to be built upon.

Nutrition surveillance has traditionally been seen as having three major purposes (Table 5-2): long-term planning, program monitoring, and timely warning (Mason et al., 1984). Timely warning of crises is the most relevant in this

TABLE 5-2 Population Data Sources and Their Use

Source	Long-Term Planning	Program Monitoring and Evaluation	Timely Warning to Prevent Crises
Repeated national surveys	Main use	Possible use, but rare as process data limited and design not ideal	No use, as too infrequent and too much lag time
Area-level surveys	Not usually, but some potential with further analysis	Possible use, but rare as process data limited, design not ideal, and external validity may be unclear	Main use, together with other data (e.g., prices)
Reporting systems	Not usually; considered less reliable than repeated national surveys	Potential use for process monitoring if lag time can be reduced	Potential main use if lag time can be minimized
Sentinel systems	Potential use	Potential use for evaluation if carefully designed	Potentially important use

context, and the tool that is particularly useful in this context for timely warning is small-scale area-level surveys. More than 1,000 small-scale surveys took place in the past 6 or 7 years in the Horn of Africa alone (Mason et al., 2007). These are useful in allowing fairly rapid assessments of the current nutritional situation and should be continued in a consistent manner.

The collective experience of reporting systems from clinics or nutrition programs in Africa is that they are not sustainable. If the reporting system is useful at the local level, then it is continued, but if people are collecting information just to pass up to other agencies, the reporting system is not continued. Again, these systems need to be part of ongoing programs and need to be linked to community-based programs that are probably the most effective way to deal with child malnutrition.

Sentinel systems (facilitated reporting from selected sites, either of data routinely collected, or by household sample surveys) are appealing because they use information from a limited number of places with implications for predicting trends more widely. Where applied, such systems have been successful. For example, in Zimbabwe, with UNICEF support, 6 monthly estimates have been made over the past 2 or 3 years, and these estimates have given remarkable information about the extent of, and changes in, malnutrition.

One outcome has been to show that the proportion of people who had one meal or less the previous day went from about 15 percent in 2007 to 34 percent in July 2008, and to 47 percent in November 2008 (*Zimbabwe Combined Micronutrient and Nutrition Surveillance Survey: Summary of Main Findings*, 2008; *Zimbabwe Nutrition Sentinel Site Surveillance System: Summary of Main Findings*, 2008).

In practice, nutrition interventions are more effective and sustainable when based upon much larger programs. For example, the health extension program in Ethiopia, supported by The World Bank, is a massive reform that increased outreach of health services. This program provides an opportunity to build on nutrition-protecting programs and make them more effective. The role of nutrition information for safety net interventions, then, is to target populations; measure their needs, the levels, and particularly the changes in nutritional status; assess the outcomes (including in terms of evaluation); and conceivably contribute to individual eligibility assessment.

Next Steps

Intervention policies of a large scale are required if such policies are to reverse the increased hunger and malnutrition that undoubtedly resulted from the food price and economic crises starting in mid-2008, even if the extent of these problems are not yet measured in any detail. The international nutrition community needs to help develop systems to provide timely and disaggregated information to support these large-scale intervention policies. Nutrition surveillance should focus more on routine data collection (e.g., prices) and qualitative assessment (e.g., household hunger scale), as well as continuing to use representative surveys. There will likely be a convergence of the safety net eligibility at the individual and population levels, as well as a convergence between the response to such crises as droughts and economic stress and the response to other large crises such as HIV/AIDS.

INSIGHTS FROM 25 YEARS OF HELEN KELLER INTERNATIONAL'S NUTRITION SURVEILLANCE IN BANGLADESH AND INDONESIA

Andrew Thorne-Lyman, M.H.S., Department of Nutrition
Harvard School of Public Health

There are few articles in the peer-reviewed literature documenting the impact of rising prices on nutritional and health outcomes in developing countries. Helen Keller International (HKI)'s Nutrition Surveillance Project (NSP) functioning in Bangladesh from 1990 to 2005 and its Indonesian Nutrition Surveillance System (NSS), which collected data from 1995 to 2005, stand out as relatively unique examples of surveillance systems that have generated analytical insights related to understanding the effects of economic crises on such nutritional outcomes as child underweight, stunting, maternal underweight, and micronutrient deficiencies (Torlesse et al., 2003; Block et al., 2004; Campbell et al., 2009a,b).

Unique Focus and Objectives of the NSP and the NSS

In many ways, the NSP and the NSS were distinct from the dominant paradigm of nutrition surveillance that was in place as of 1990 when the NSP was started (Bloem et al., 2008). At the time, surveillance was thought of as something that should be simple and inexpensive, including only a minimal number of indicators. Surveillance systems were intended to generate information to be used primarily at the community level to influence programs. The NSP and the NSS diverged from this model. From inception, both systems were oriented towards generating information that would be used to inform decision making at the national level, by policy makers working in multiple sectors including nutrition, health, and agriculture.

Design of the Systems

Nutritional status is an outcome that is the result of many complex causes and interactions. Analysis of such interactions through conventional nutrition surveillance approaches is often limited by both the sample size of surveillance systems and the relatively narrow scope of other variables (which typically reflect the sector providing funding to the system). Both the NSP and the NSS had relatively large sample sizes that enabled both geographic disaggregation and

TABLE 5-3 Comparison of the NSP in Bangladesh and the NSS in Indonesia

	NSP (Bangladesh)	NSS (Indonesia)
Implemented by	HKI and GOB (IPHN); network of 17 NGO partners	HKI
Dates of surveillance	1990–2005	1995–2005
Frequency of data collection	6 times/year	4 times/year
Sampling	Multistage cluster	Multistage cluster
Sample size per round	≈10,490 rural ≈1,300 urban	≈33,600 rural ≈10,800 urban
Statistically representative (rural)	National, divisional levels	Seven densely populated provinces (70% of rural population)
Urban	Slums in three cities	Slums in four large cities

NOTE: GOB = government of Bangladesh; HKI = Helen Keller International; IPHN = International Poverty and Health Network; NGO = nongovernmental organization.

precise estimation of such "tip of the iceberg" indicators as child night blindness, as well as facilitating disaggregated analyses that might not have otherwise been possible (Table 5-3).

The NSS in Indonesia began in Java and expanded when the economic crisis hit, tracking the impact of the crisis and progress made as the country emerged from that crisis. In contrast, the system in Bangladesh was more oriented to collecting data to track development programs and to track the impact of development on a national level, although it was also used to measure the impact of crises such as hurricanes, pests, and flooding that affect different parts of the country over time.

Malnutrition is a seasonal phenomenon in many countries, and one of the benefits of an ongoing surveillance system that collects data at multiple points throughout the year is the ability to understand the normal seasonal patterns of malnutrition, whereas a stand-alone survey taken at one point in time could easily reach the wrong conclusions about trends in malnutrition or might wrongly attribute a rise or fall in malnutrition to a particular event (which might just be a regular seasonal phenomenon).

Another unique feature of the NSP and the NSS was the inclusion of urban slums in the samples in both Indonesia and Bangladesh, which enabled the system to provide information about a growing segment of urban poor populations in both countries. The use of modules in the design of the system enabled flexibility to collect information about emerging issues of programmatic relevance, such as reasons for lower coverage of the vitamin A program in Chittagong, cyclone preparedness, and variables to be used to help measure the impact of the growing national nutrition program.

Quality Control

Quality control was an essential part of both the NSS and the NSP, especially because information was collected from a network of more than 17 partner organizations. In Bangladesh—a country prone to floods, cyclones, and droughts—having a network of different sites throughout the country was invaluable during times of crisis. One of the downsides of using many partners, however, is the variability in the quality of the data collected. In an effort to ensure good quality across all of the partners, there was a quality control team that revisited 10 percent of the households, administered part of the questionnaire, and made sure the questions were answered in the same way as had been previously reported. Refresher trainings were also held prior to each round of data collection to ensure good quality data.

Breadth of Information Collected

Because of the breadth of information collected, the NSP and the NSS were able to not only show trends in malnutrition and stunting over time, but also to explain the factors associated with these changes. Although information about the prevalence of malnutrition is often collected every several years in many countries through other surveys, one of the virtues of the NSP and the NSS was the inclusion of information not typically available through health-focused surveys, such as food prices, household expenditures, agricultural and cropping patterns, land ownership, and female decision making within the household. These enabled a wider exploration of factors associated with malnutrition than is normally undertaken, as well as the isolation of relationships between variables by enabling adjustments for confounding factors. In both countries, information on micronutrient status was also collected, facilitating the understanding that changes in the quality of the diet resulting from food price changes may have adverse impacts on health, even if they do not influence the prevalence of child undernutrition in affected populations (Bloem et al., 2005).

Weaknesses of the NSP and the NSS Systems

Often, surveillance systems are expected to serve as early warning systems. This was not an explicit goal of either the NSP or the NSS. The relatively large sample size required collection of data over a period ranging from 4 to 6 weeks—such turnaround time makes it difficult for information to be used as an early warning. One downside of most early warning systems is that their reliance on small sample sizes leads to a lack of precision. Differences in the objectives of surveillance systems should influence their design.

The issue of cost and sustainability is another criticism that has been made of these systems. Although cost information is not readily available, these systems were more expensive than most. But in the context of the present food crisis, many rapid assessments were done at significant cost, because they required the establishment of new infrastructure to collect information. Most lacked baseline information from the time period prior to the food price rises that could have been used to better understand the potential impacts of the crisis itself. The value of having ongoing systems in place to collect information is often not appreciated until after a crisis hits.

FAMINE EARLY WARNING SYSTEMS NETWORK, NUTRITION SURVEILLANCE, AND EARLY WARNING

Chris Hillbruner, M.S., Food Security Warning Specialist
Chemonics

Introduction to FEWS NET

The goal of the Famine Early Warning Systems Network (FEWS NET) is to provide an early warning of an impending food crisis. While FEWS NET is not a direct implementer of nutrition surveillance systems, it has a role to play in collecting nutrition information that could provide an early warning of food insecurity and other crises that are detrimental to the nutritional status of surveilled populations.

FEWS NET is at work in 20 countries and has broad experience looking at the type of information available in different places and applying that data where it can be successfully used. FEWS NET is a partnership that pulls together a variety of different information flows from a core group of international partners, including academic institutions, consulting companies, and U.S. government institutions. FEWS NET also collaborates with a wide network of formal and informal partners. FEWS NET has nine internal partners: Chemonics International, Michigan State University, FEG Consulting, WebFirst, Inc., Intana, U.S. Geological Survey, National Oceanic and Atmospheric Administration, U.S. Department of Agriculture, and National Aeronautics and Space Administration. Members and collaborators of FEWS NET include UN World Food Programme (WFP); FAO; several national ministries of agriculture, rural development, and health; various price and market information systems; meteorological centers; nongovernmental organizations (NGOs); and other UN agencies (UNICEF, UN High Commissioner for Refugees [UNHCR]).

FEWS NET focuses primarily on early warning with a 4- to 6-month outlook, providing information that is credible and actionable for decision makers.

The system also works toward building the capacity of national early warning systems. The 20 countries covered by FEWS NET are primarily in sub-Saharan Africa, but also include Haiti, Afghanistan, and Guatemala. In each of these countries, FEWS NET attempts to provide an integrated food security analysis including agro-climatic information, crop forecasts, food prices, and, when possible, nutritional information. This information is provided by a variety of different sources and is evaluated in a "livelihoods framework" that allows a baseline understanding of how people earn their livelihood in a normal year. The additional information gained through continual surveillance is used to make predictions about food security conditions in the future.

A Potential Role for Nutrition Information in FEWS NET Analysis and Reporting

There are four main areas where there is a potential role for nutrition information in FEWS NET's work. The first is vulnerability analysis—looking at how underlying levels of malnutrition in a population make them more vulnerable to climate changes, conflict, or other phenomena. Second, nutrition information can play a role in early warning. This is a somewhat controversial perspective, but there is a case to be made. Third, nutrition information can be used in crisis monitoring. In many FEWS NET countries (although the goal of the system is to provide early warning), once a crisis has begun, FEWS NET provides ongoing monitoring. In this way, nutrition information can play an important role in crisis monitoring as well. Finally, it must be noted that nutrition information is useful in advocacy. No matter how good all other information may be, there is a resonance that nutrition data has with decision makers that makes it a very useful tool in trying to convey the gravity of the situation. For this reason, the international nutrition community must be extraordinarily careful about the nutrition information it uses, because it can have an important effect on policy.

Challenges in Incorporating Nutrition into Food Security Early Warning

Ideally, FEWS NET would use nutrition information as described above. The reality is that nutrition information is incorporated very little in much of the analysis that FEWS NET does.

The first reason for this lack of incorporation is that nutrition information is not commonly viewed as having a role to play in early warning. Nutrition is often considered a lagging indicator that has nothing to contribute to early warning; once malnutrition is detected, it is too late. But using nutrition information in early warning is much more nuanced than most realize. For example, "nutrition data" can mean many different things—a variety of indicators, different collection methods, and different geographic coverage. For example, while a national-level estimate of acute malnutrition that is collected once per year is probably not particularly useful for early warning, surveillance data collected in vulnerable areas could be quite valuable, especially if it is collected over time, allowing for comparisons between monthly data and seasonal averages. There is a real need for the nutrition, food security, and early warning communities to start thinking a bit outside the box about how they can do a better job of incorporating this information into early warning analysis and reporting.

Timeliness and regularity of nutrition information flows are a second challenge to incorporating nutrition data into food security early warning. Even when systems are in place, if the data are not timely and are not reported regularly, they become of limited use for early warning.

Third, obtaining complete high-quality data can be challenging. Some of the nutrition information that is collected contains many methodological shortcomings. In other cases, the quality of anthropometric information may be excellent, but no other information was collected. In such cases, information on very high wasting rates in a certain area may be available, but there is no contextual information to give a sense of what is actually going on and what kind of response should be initiated.

Examples of Nutrition Information Systems and Nutrition Data Flows in FEWS NET Countries

Niger has a national-level system that is run by the government. The nutrition information they have been collecting monthly since 2006 includes admission to feeding centers, disease incidence, mortality rates, and case fatality rates, which offer some contextual information about health. FEWS NET integrates much of the nutrition data from Niger into its analysis. One problem, though, is that the admissions data are confounded by the fact that the number of feeding centers in Niger is constantly changing, which complicates looking at trends over time.

In Kenya, the Arid Lands Program collects data across two-thirds of the country and publishes monthly reports for each of about 30 districts where the system operates. The Arid Lands Program uses a somewhat unique indicator—the percentage of children with a mid-upper arm circumference (MUAC) less than 135 mm—and this can make it difficult when using the data with decision makers who are more familiar with traditional nutrition indicators. The Arid Lands Program also collects some contextual information such as food and livestock prices and rainfall. This system has been collecting data long enough that an examination of trends and comparisons with seasonal averages is possible. Unfortunately, the analytical capacity of the system is somewhat limited.

In Somalia, the Food Security and Nutrition Analysis Unit performs nutrition surveys in different parts of the country during the year. In terms of data quality, depth, and reliability, they are probably the strongest source of nutrition information that FEWS NET has. However, again from an early warning perspective, because the data collection is accomplished through large-scale surveys rather than regular surveillance, it is more difficult to look at trend analysis and use that information to project into the future.

Moving Forward

There is a need to start advocating for a bigger role for nutrition and health data in early warning systems. Additionally, *more* nutrition surveillance systems must be developed. These should be timely and ongoing, focus on vulnerable areas and populations, have some analytical capacity (not just a data collection function), and leverage partnerships for contextual data collection. A system should recognize who is currently collecting information and focus on where the gaps are, as opposed to creating a heavy and duplicative system.

LISTENING POSTS PROJECT: A CONCEPT FOR A REAL-TIME SURVEILLANCE SYSTEM NESTED WITHIN A PROGRAM

Anna Taylor, Head of Hunger Reduction
Save the Children UK

Introduction to Listening Posts

Save the Children and Action Against Hunger are collaborating in an effort to integrate surveillance into routine program work. During the food price increases of 2008, Save the Children realized it was ill-prepared to determine the effects of the crisis in the communities where it had programs. In an effort to rectify this gap, the Listening Posts project was developed; it has a threefold purpose:

- Listen and pick up shocks and their nutritional effects at a local level.
- Respond and be able to use information to inform program work and adjust programming according to the findings of the surveillance.
- Inform others about what is happening in the vulnerable communities under the project's surveillance, and link what is happening at the local level to global shocks as they occur, in hopes of implementing a larger response.

The following six principles guide the design of the Listening Posts system:

1. The system should be as light as possible so as to be easily integrated into program work with very little additional cost. The proposed level of investment is one team of two people working for 5 days every 3 months gathering data.

2. The system should operate as rapidly as possible. The project is currently investigating new technologies for data management, such as mobile technology, in order to transmit data quickly from local points to capital entities and eventually to the rest of the globe.
3. The system should be local and global. The project is attempting to strike a balance between information that is *locally useful and context specific*, as well as information that allows discussion with some authority about what is happening to vulnerable communities in multiple locations for *global policy application*.
4. The system should be as widely endorsed as possible to ensure the findings are quickly accepted and acted upon.
5. The system should be linked into national early warning systems. Ideally, the data would be useful for FEWS NET and used in the food security classification processes that are happening increasingly in various countries.
6. The system should be as replicable as possible so that, if successful, it can be easily used by other programs.

Indicators to Be Collected

The Listening Posts project would use a very small indicator set. It would collect data on the prices of staples and the ratio between those prices and labor rates as a very broad measure of changes in the economic situation. Dietary diversity and feeding frequency among children aged 6 to 24 months would also be collected using the standard indicators, along with mean weight gain, MUAC, and edema for children aged 6 to 24 months.

Some optional indicators that country programs may choose to use include details around coping strategies and talking to children about their experiences as well as such impact-level assessments as feeding center admissions, changes in child labor, marriage, and reduced school attendance. Such data would be more relevant, for example, when an education program is running in the same area.

How Would It Work?

Country programs would be required to implement the Listening Post strategy in a minimum of one livelihood zone. Each livelihood zone would be divided into six quadrants, selecting the community closest to the center of the quadrant for the nutritional data; that community would be the listening post. There would be six listening posts per livelihood zone. At each listening post, 16 children would be selected to survey. When a child exceeded 24 months of age, a new child would be selected. According to this design, there would be 96 children per livelihood zone.

What Is the Cost-of-the-Diet (CoD) Assessment?

The cost-of-the-diet (CoD) assessment developed by Save the Children over the past few years is a method for calculating the lowest cost diet that meets the nutritional requirements of a whole family. It builds on the discussion about data quality and the international nutrition community's inability to talk about the affordability of a quality diet. Food security is often inappropriately equated with energy security; the CoD assessment is an attempt to take the discussion a step further toward understanding affordability of a quality diet.

CoD is an Excel-based tool that originates from the linear programming work that has been done at the World Health Organization (WHO). The system requires a list of locally available foods, their prices by season, and the size of a typical family. That data are entered into the linear programming tool, which then calculates the lowest cost nutritionally adequate diet for that location (Figure 5-2). This unique assessment approach takes seasonality into account and determines locally specific estimates of cost. It is also possible to build constraints into the program to limit the portion sizes of very cheap foods, which may not be possible to eat in large quantities, and to model the effects of introducing micronutrient supplementation or home fortification.

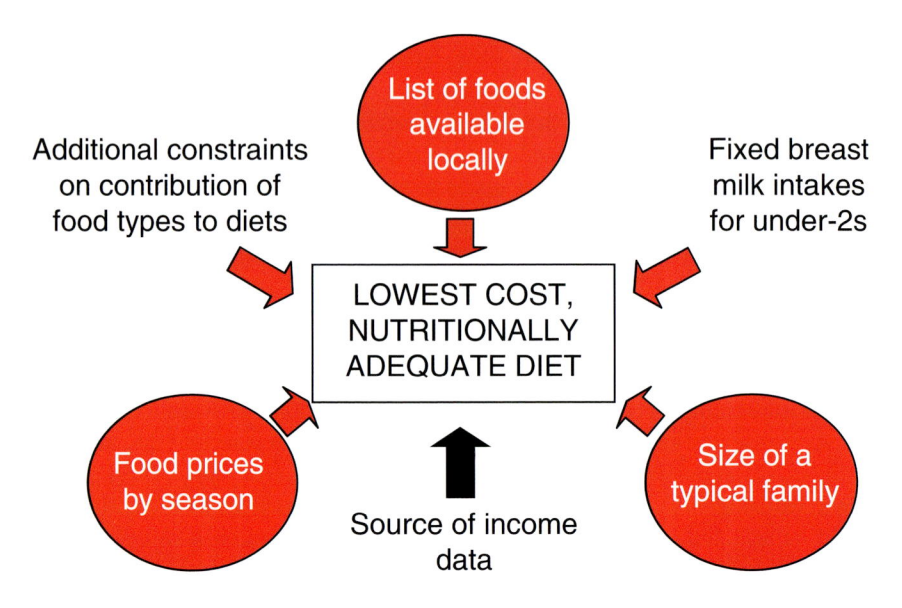

FIGURE 5-2 Schematic of the cost-of-the-diet assessment.

Country Example

From 2005 to 2006, the CoD system was used in Bangladesh. It found that in one community, 194 households, or 50 percent of families, could not afford a diet that met all nutritional requirements. (This scenario was estimated even before non-food items that the households may want to buy were considered.) About 15 percent of households could not even afford an energy-only diet (meeting only basic energy requirements and not other nutrient needs).

In 2007–2008, a repeat assessment looked at the nutritional effects of food price increases. The system found that when considering both rising food prices and the harvest failure, there was not much difference in the proportion of people who could afford a quality diet. In fact, this second assessment showed that those who could afford a quality diet *slightly* increased, although the proportion who were pushed further into poverty and were unable to even afford an energy-only diet *massively* increased. Save the Children UK will use CoD data as a key element of the Listening Posts surveillance system and will launch the software and guidelines in Rome in October 2009.

Goals for the Future

Ideally, the Listening Posts project is anticipated to involve quarterly data collection for the entire Listening Post system. One or two livelihood zones per country would be covered in this quarterly reporting, and all of the data would be fed into national early warning systems. Ultimately, an annual global report, capturing findings from a number of livelihood zones over a number of countries, could be produced on a quarterly basis.

FOOD SECURITY, NUTRITION MONITORING, AND THE GLOBAL FOOD PRICE CRISIS: USAID/FFP TITLE II PROGRAMS

Ellen Mathys, M.P.H., Senior Food Security
Early Warning and Response Specialist
Food and Nutrition Technical Assistance Project II (FANTA-2)
Academy for Educational Development

FANTA-2 and Food Security Programs

The U.S. Agency for International Development (USAID)-funded Food and Nutrition Technical Assistance Project II (FANTA-2) undertakes activities designed to strengthen the capacity of awardees that are implementing multiyear development projects funded with USAID Office of Food for Peace (FFP) Title II resources (MYAPs[2]). FANTA-2 support to these projects includes technical support, support for monitoring and evaluation, and in some cases support to nutrition surveillance activities. These awardees collect and report on program monitoring and evaluation data to the FFP, and the opportunity exists to use these MYAP data to monitor the food security and nutrition effects of the global economic and food price crises. These data that are already being collected for annual program monitoring and impact evaluation may be used to understand how chronically food-insecure communities are affected by these price shocks.

FFP Requirements: Impact Indicators

FFP Title II awardees are required to report on the following impact-level indicators for program evaluation, which allows the FFP and FANTA-2 to look at the following population-level outcomes in program areas:

- Average number of months of adequate household food provisioning
- Average household dietary diversity score
- Percentage of children aged 0–59 months who are underweight
- Percentage of children aged 6–59 months who are stunted

A typical MYAP is 5 years long, and impact-level indicators are measured during the baseline study and final evaluation. These impact-level indicators are fixed (i.e., the indicator definitions and data collection techniques are standardized). This limits how the data can be used to understand such things as seasonality because it is only assessed at baseline and final evaluation. However, the standardization of the indicators allows comparability when compiling data globally.

FFP Requirements: Monitoring Indicators

Title II awardees are also required to collect data on four program monitoring indicators annually. First, an anthropometric indicator of the MYAP's choice must be used to regularly monitor the maintenance or improvement in nutritional status of beneficiaries. Such an indicator must reflect anthropometric measurements of child or adult growth, or graduation based on anthropometric measures. Acceptable indicators include the prevalence of stunting, underweight, wasting, low body mass index (BMI), or low MUAC; trends (weight gain, growth faltering); and graduation or exiting—the proportion of children or adults recuperating, according to defined anthropometric cutoffs. Second, the percentage of benefi-

---

[2] Throughout this presentation, these Title II–supported multiyear assistance programs will be referred to as MYAPs.

ciaries adopting improved health, nutrition, or hygiene behaviors is measured. Third, the percentage of beneficiaries (in this case, farmers) using a project-defined minimum number of sustainable agricultural technologies is recorded. Fourth, the MYAP must determine the number of project-assisted communities with (due to the MYAP) improved physical infrastructure to mitigate the impact of price shocks, disaster early warning and response systems in place, safety nets to address the needs of the most vulnerable members, and improved community capacity.

FFP Guidance: Trigger Indicators

Trigger indicators refer to early warning indicators that MYAPs are encouraged, but not required, to collect. Trigger indicators are used to determine the threshold at which MYAPs need to shift activities or when additional resources for new activities are required in response to a slow-onset shock. Trigger indicators typically include such variables as:

- Climate or rainfall,
- Key food and cash crop production,
- Staple prices,
- Livestock prices,
- Coping strategies,
- Remittances and debt, and
- Nutrition trends (if available).

MYAPs identify trigger indicators in advance and can collect them in an ongoing manner either through secondary data sources or primary data collection activities. On the basis of trigger indicator information, MYAPs can request emergency resources for use in their MYAP area. This serves as an administrative convenience in a way that allows them to protect the gains of the MYAP while they respond to a deterioration of food security conditions in the area.

Many NGOs do not feel comfortable performing what they see as early warning activities; however, the benefits are savings in time, money, and the ability to respond quickly on the basis of localized information without the procedural requirements previously associated with emergency response (i.e., through development of a single-year assistance program proposal).

Availability of Population-Level Outcome Data in MYAP Areas

Although the FFP implements food security programs in many of the world's most chronically food-insecure communities, the opportunity to use data collected on these populations—especially nutritional data—is not being used to full advantage. In 2007 for example, only 31 of the 79 MYAPs were in priority FFP areas and reported underweight data. Of these 31 MYAPs, only 7 were reporting population-representative prevalence rates (the data were collected via the MYAP's baseline or final evaluation). Thus only a minority of programs are providing population-representative outcome data useful for nutrition surveillance. This is unfortunate because these MYAPs are typically in the most chronically food-insecure areas in their countries.

Conclusions and Key Opportunities

There are both positive trends and challenges that must be noted in a discussion of Title II awardees and their MYAPs. In terms of positive trends, there has been an increasing availability and quality of food security and nutrition data collected by MYAPs and their key national and international partners. Additionally, there has been strong support by USAID for monitoring and evaluation (including trigger indicators) in Title II programs. In terms of challenges, there are not enough resources to increase monitoring and evaluation responsibilities for MYAPs; yet the quality, scale, and frequency of data collection—particularly of population-representative data—are insufficient. It is difficult to compare the anthropometric monitoring data among MYAPs because of the variation in indicators used, the methodological constraints to attributing causality, and the complicated task of defining and harmonizing trigger indicators. Clearly, there is a role for nutrition surveillance systems to monitor impacts of price shocks on food security and nutrition, and this will complement, not duplicate, data collected through MYAPs.

There are some key opportunities to improve current data collection and use of the data that are collected through MYAPs. In terms of data collection, several program approaches on the horizon would expand nutrition surveillance coverage, including the prevention of malnutrition in children under 2 years and mass treatment. New food security indicators are being developed that can be easily integrated into household surveys and monitoring, including the household hunger scale. Efforts are underway to strengthen and harmonize trigger indicators for food security monitoring and early warning in MYAPs. Mobile phone technologies are being piloted for real-time transmission of food security early warning data. NGOs increasingly collaborate in national food security and nutrition monitoring and early warning networks. In terms of data analysis and data use, new technologies are being developed to link multiple, agency-specific data sets together to enable triangulation and foster coordination. Food security scenario development must occur (including contingency and response planning) and a link must be forged between national and international alert systems.

DISCUSSION

This discussion section encompasses the question-and-answer sessions that followed the presentations summarized in this chapter. Workshop participants' questions and comments have been consolidated under general headings.

Which Indicators?

Several workshop participants argued that surveillance systems should be collecting very simple and interpretable indicators, like the cost of a local food basket or number of hours worked. The latter tracks the wage rate, a very easily understood indicator that is not very accurate, but says something that people understand. Such interpretable indicators track both wages *and* food prices, which are both extremely useful in this context. On the other side of the spectrum, such indicators as, "How many times a week do you eat a vitamin A-containing food?" is too complicated. Not only is the answer to such a question difficult to estimate, but the results are equally complex to interpret.

This session on nutrition surveillance illustrated the range of data collection and systems that often function in the same country, for different organizations, using different types of monitoring systems. One participant suggested that when these data are brought back to counterparts at the national level, it can be confusing and less than helpful. It is urgent that the international nutrition community agree upon a set of indicators and approaches for certain conditions and purposes. It would do a great deal of good to have a sense of unity around interventions and indicators for nutrition. Because at the moment there is a greater sense of agreement on a set of interventions that have proven to be effective for nutrition, the nutrition community should be vocal about what interventions it supports and what indicators it wishes to collect.

It was also posited that while coming up with a set of common indicators may be useful, it should also be noted that there are places where no data are being collected at all. In such countries, the need to start collecting *some* level of information may be more important than agreeing upon a certain set of data.

The Time to Intervene

One participant urged the international nutrition community to use the food basket (that corresponds to the culture and the price of that food basket relative to income) as the indicator that determines whether people have sufficient food quantity or quality. There needs to be more thinking in terms of region and ethnic group because indicators change according to these variables. WHO should discard set cut-points for when agencies react. For example, if a country shows 14.8 percent wasting (with the WHO cut-point at 15 percent); is it logical to say that only if wasting increases another .2 percent will there be a crisis? Additionally, the wasting indicator is inappropriate. For example, there was no increased wasting at all in sub-Saharan Africa during the HIV/AIDS crisis in the early 2000s, but there was a huge increase in underweight. These set indicators and fixed cut-points can easily be misinterpreted. A workshop participant urged that, instead, surveillance systems should collect situation-specific data and undertake a more sophisticated interpretation than merely stating that 15 percent wasting signifies a crisis.

Ability of Surveillance Systems to Accurately Predict

One workshop participant expressed concern that the trigger that exists in surveillance systems presumes that there is a high level of predictive value. Given that crises are rare (even though epidemiologically high sensitivity and specificity in terms of separating crisis from noncrisis is possible), the ability to predict crises with high levels of certainty using data from surveillance systems may be weak. Dr. Mason responded that there are enough crises to make predictions; even historical data sets can be used. For example, the FAO early warning system accurately predicts food crises. But it is remarkable how seldom surveillance systems are used for early warning of food and nutrition crises.

REFERENCES

Block, S. A., L. Kiess, P. Webb, S. Kosen, R. Moench-Pfanner, M. W. Bloem, and C. P. Timmer. 2004. Macro shocks and micro outcomes: Child nutrition during Indonesia's crisis. *Economics and Human Biology* 21(1):21-44.

Bloem, M. W., S. de Pee, and I. Darnton-Hill. 2005. Micronutrient deficiencies and maternal thinness: First link in the chain of nutritional and health events in economic crises. In *Primary and Secondary Nutrition* 2nd ed. Edited by A. Bendich and R. J. Deckelbaum. Totowa, NJ: Humana Press. Pp. 357-373.

Bloem, M. W., S. de Pee, and R. D. Semba. 2008. How much do data influence programs for health and nutrition? Experience from health and nutrition surveillance systems. In *Nutrition and Health in Developing Countries.* 2nd ed. Edited by R. D. Semba and M. W. Bloem. Totowa, NJ: Humana Press. Pp. 831-858.

Campbell, A. A., de Pee S., Sun K., et al. 2009a. Relationship of household food insecurity to neonatal, infant, and under-five child mortality among families in rural Indonesia. *Food and Nutrition Bulletin* 30(2):112-119.

Campbell, A. A., A. Thorne-Lyman, K. Sun, S. de Pee, K. Kraemer, R. Moench-Pfanner, M. Sari, N. Akhter, M. W. Bloem, and R. D. Semba. 2009b. Indonesian women of childbearing age are at greater risk of clinical vitamin A deficiency in families that spend more on rice and less on fruits/vegetables and animal-based foods. *Nutrition Research* 29(2):75-81.

Deitchler, M., T. Ballard, A. Swindale, and J. Coates. 2009. *HFIAS Validation Study: Identifying an Experience-Based Measure of Household Hunger for Cross-cultural Use.* Washington, DC: Academy for Educational Development.

FAO/WHO/UNICEF. 1976. *Methodology of Nutritional Surveillance.* Geneva: World Health Organization.

Food and Agriculture Organization of the United Nations. 2002. *Proceedings of International Scientific Conference: Measuring and Assessment of Food Deprivation and Undernutrition, 2003.* Rome: Food and Agriculture Organization.

———. 2009. *More People Than Ever Are Victims of Hunger.* Rome: Food and Agriculture Organization.

Mason, J., J.-P. Habicht, H. Tabatabai, and V. Valverde. 1984. *Nutritional Surveillance.* Geneva.

Mason, J., et al. 2007 (unpublished). *The Impact of Drought and HIV on Child Nutrition in Eastern and Southern Africa.* Tulane University.

Social Safety Nets: Moving Forward for an African Agenda. 2009 (unpublished). The World Bank.

Torlesse, H., L. Kiess, and M. W. Bloem. 2003. Association of household rice expenditure with child nutritional status indicates a role for macroeconomic food policy in combating malnutrition. *Journal of Nutrition* 133(5):1320-1325.

UN ACC-SCN. 1993. Second report on the world nutrition situation. *UN-SCN* II:111-114.

Zimbabwe Combined Micronutrient and Nutrition Surveillance Survey: Summary of Main Findings. 2008. GoZ Food and Nutrition Council and UNICEF.

Zimbabwe Nutrition Sentinel Site Surveillance System: Summary of Main Findings. 2008. GoZ Food and Nutrition Council and UNICEF.

6

The Global Response to the Crises

A wide constellation of people and organizations work on nutrition and food security issues. These include, for example, multilateral United Nations (UN) agencies, bilateral government agencies, nongovernmental organizations, universities, research institutions, foundations, and the private sector. As described by moderator Hans Herren of the Millennium Institute, the following presentations helped workshop participants understand the landscape of the global nutrition field, the people and organizations who work in it, and their respective roles, functions, and capacities to respond to the outcomes of the recent food price and economic crises.

INTRODUCTION TO THE GLOBAL NUTRITION LANDSCAPE

Ruth Levine, Ph.D.,
Vice President for Programs and Operations; Senior Fellow
Center for Global Development

The Center for Global Development recently published a document titled *Global Nutrition Institutions: Is There an Appetite for Change?* The project followed work done for the *Lancet* series on maternal and child undernutrition, in which some key weaknesses were identified in the way international institutions are organized and work together, and the level of funding and capacity they have (Morris et al., 2008). The paper took a closer look at these issues with the hope of provoking thinking and conversation about how the institutional arrangements could be improved to serve the current needs of the international nutrition community (Levine and Kuczynski, 2009).

Global Nutrition Landscape Comprised of Many Actors, Spread Thinly

A large set of different roles needs to be filled in any sector or across any sectors, and the nutrition and food security areas are no exception (Table 6-1). Owing to historical and other factors, there is persistent confusion and sometimes conflict around which institutions play which roles and how they do that together.

One role could be characterized as that of setting norms. In this area, the World Health Organization undertakes work, and the UN Standing Committee on Nutrition also acts as a forum for UN agencies to discuss topics on this subject.

Research activity is principally being undertaken by academic institutions and in specialized research institutions. Generally, research is quite widely disseminated but often without a defined role to play in policy and programmatic decisions.

TABLE 6-1 Illustrative Organizations Active in International Nutrition

Category	Organization	Key Role(s) Related to Nutrition
Multilateral Agencies		
	UNICEF	Program implementation focused on maternal and child health, norms, and standard setting. Focus is on nutrition security, micronutrients, breast-feeding, and emergency response.
	United Nations Standing Committee on Nutrition	Network of food and nutrition professionals. Promotes cooperation among UN agencies and partner organizations, including NGOs, in support of community, national, regional, and international efforts to end malnutrition.
	The World Bank	Project and sector financing to countries with loans on near commercial and soft terms. Supports government implementation of projects and policy reforms with technical assistance from The World Bank staff and consultants.
	World Food Programme (WFP)	Implementation of emergency response and food aid. Provides logistics and support through development programs. Operates school feeding programs.

TABLE 6-1 Continued

Category	Organization	Key Role(s) Related to Nutrition
	World Health Organization (WHO)	Sets standards, and establishes policies and programs. Biomedical and public health focus on reduction of micronutrient malnutrition, growth assessment, and surveillance.
Bilateral Agencies	Canadian International Development Agency (CIDA)	Donor with focus on micronutrient and other technical interventions, such as vitamin A programming and iodine.
	U.S. Agency for International Development (USAID)	Largest bilateral donor; focus on targeted maternal and child health projects, micronutrient interventions.
Nongovernmental Organizations	Academy for International Development	Short-term technical assistance, product research and marketing.
	CARE	Technical support. Focus on the delivery of food commodities and resources during emergencies.
	Global Alliance for Improved Nutrition	Supports public-private partnerships to address micronutrient deficiencies.
	Helen Keller International	Intervention delivery; research, and advocacy functions. Focus on nutrition, child survival, and eye health.
	Manoff Group	Intervention delivery, communications and behavior-centered programming.
	Micronutrient Initiative	Intervention delivery and research. Focus on micronutrient and vitamin deficiencies, vitamin A supplements, fortification.
	PATH	Development of new diagnostics for micronutrient deficiencies; innovation in biofortified foods.
Universities and Research Institutions	Consultative Group on International Agricultural Research (CGIAR)	Research; alliance of members, partners and 15 international agricultural centers. Focus on food security.

TABLE 6-1 Continued

Category	Organization	Key Role(s) Related to Nutrition
	Cornell University	Training and research, including basic science, community nutrition, policy development.
	Instituto de Investigación Nutricional (Lima, Peru)	Research and program implementation, teaching and training services in health and nutrition. Focus on community health in Peru.
	International Center for Tropical Agriculture (CIAT)	Development of biofortified foods.
	International Food Policy Research Institute (IFPRI)	Scientific research and related activities; supported by CGIAR; focus on food security and poverty reduction. Implementing HarvestPlus.
	Johns Hopkins University Bloomberg School of Public Health	Research and training in public health nutrition.
	London School of Hygiene and Tropical Medicine	Research and training in public health nutrition.
	Mahidol University, Thailand	Research and policy analysis, training, and consultation.
Private Sector		
	Danone	Grameen Danone partnerships to promote local entrepreneurship in nutrition.
	Unilever	Partnership with the WFP to improve the nutrition and health of poor school-aged children.
Philanthropies		
	Bill & Melinda Gates Foundation	Focus is on reducing micronutrient deficiencies and undernutrition in vulnerable groups, particularly women and children less than 2 years through Global Health Program; Global Development Program includes grant making to increase quantity and quality of staple foods.
	Children's Investment Fund	Emerging emphasis on nutrition and food security as part of long-term development programs.

SOURCE: Levine and Kuczynski, 2009.

A small number of funders are active in international nutrition. The U.S. Agency for International Development (USAID) and the Canadian International Development Agency have been longstanding financial supporters among the development agencies. There are now some new entrants into funding nutrition and food security efforts, including Irish Aid, the United Kingdom (UK) Department for International Development, the European Union (EU), and the Bill & Melinda Gates Foundation.

On the implementing side, there are large international nongovernmental organizations (NGOs) such as Save the Children and CARE. Important implementers are also governments in developing countries where, within ministries of health, social protection, and to some extent ministries of agriculture, there are nutrition-related activities underway, funded either through domestic sources or with external support.

Several entities are involved in policy development, advocacy, and building capacity. The major UN agencies—including the United Nations Children Fund (UNICEF), the World Health Organization (WHO), and the World Food Programme (WFP)—all play roles across many of those different domains: implementation, research, generating and allocating funding, setting norms, and building capacity.

Major Challenges

Three inherent characteristics of the field of nutrition make it a special challenge institutionally, as well as in terms of communication to policy makers. First, there is not always a clear, direct link between nutritional status and tangible, measurable outcomes such as lives lost. As a result, the international nutrition community ends up being dependent on disability-adjusted life years (DALYs), which are not very compelling or intuitively understood. Further, it is not entirely clear under what health problem (or even social problem) nutrition broadly fits, because poor nutrition contributes to many types of poor health outcomes; in this way, nutrition is everyone's problem, but no one's responsibility. In addition, there are quite profound measurement errors that interfere with the ability of institutions and policy makers to determine what to do about nutrition. Both under- and overnutrition are relative constructs without necessarily a very direct line to a biomedical result. To motivate action and attract champions and clear lines of action, this lack of a direct link to a tangible outcome is a major challenge for global nutrition actors.

Second, nutrition is part of the poverty and social justice agenda. It is more about the distribution of resources in a society than anything else. As a result, it is not entirely clear who the nutrition "constituency" is. Relative to the size of the problem and compared to some other problems for which biomedical solutions (vaccines, drugs) dominate, there are few technological solutions. Nutrition has to do with the social determinants of health as well as a set of individual behaviors, which complicates categorizing nutrition as a specific, tractable health issue in need of a solution.

The third inherent characteristic of nutrition that makes it institutionally complicated is how profoundly it spans sectors—primarily health and agriculture. These two sectors, unfortunately, do not have a long track record of successfully working together or having a mutual understanding.

The current arrangement of institutions, and the interactions among them, does not seem able to handle either the broad nutrition challenges that have persisted for decades or the more immediate challenge of food security amidst the economic crisis. Funders encounter a fractious environment of institutions and groups attempting to establish norms and priorities and advocating their own technical solutions or implementation approaches. Groups with particular interests defend their territory, which is not surprising in an environment of restrained resources, but this undermines other messages about priorities. Partially as a result, decision makers at the national and subnational levels often receive contradictory or confusing guidance from international partners and face a somewhat chaotic set of program ideas and implementations to support.

Does the Nutrition Community Have What It Takes to Respond to the Current Challenges?

Six elements are commonly found among successful large-scale public health achievements (Levine, 2007). One is *adequate, reliable, and long-term funding*. This is clearly missing from the nutrition sector relative to the size of the problem. Funding for nutrition programs is low. Relatively few international agencies have made significant commitments at the national level. With some exceptions, domestic resources for nutrition programs are low.

The second element that is essential for success is the *existence of genuine champions and leadership* at the international and national levels. Such champions for nutrition are also missing at the moment. Nutrition is not very prominent on the global health agenda. Those in the nutrition community have a track record of talking *within* the nutrition community; they gravitate toward technical subjects and have difficulty broadly communicating consistent messages, not only about the importance of under- and overnutrition, but also about a set of actions that with sufficient resources could make a major positive difference.

The third element necessary for success is *technological innovation within effective delivery systems*. That does not mean a magic bullet. What has been observed in other global health successes is that there is a real momentum established when a new approach or new idea is "packaged" and talked about in a coherent way. There is a lot of promise within the nutrition community, and there are some excellent examples from the past, but much debate remains about the most promising approaches, especially regarding the appropriate role of the private sector as a delivery channel and as a partner in addressing nutrition challenges.

The fourth element is *technical consensus among the recognized expert community*, including both at the national level and internationally, about the magnitude of the problem and the way forward. Much progress has occurred in nutrition in the past few years, with the *Lancet* series in particular providing clear evidence and a synthesis of evidence about the window of opportunity from birth to 2 years. The series identified a set of interventions that has now been reasonably well evaluated (Bhutta et al., 2008). At the same time, there is still some debate around that technical consensus; there are remaining questions about ready-to-use therapeutic feeding, for example. There is little agreement about what the connections should (or can) be between the health, nutrition, and agricultural communities and their interventions.

The fifth element of success is *good management* on the ground, often termed *capacity*. Currently, major capacity gaps are observed in nutrition. This lack of capacity can be attributed to the persistent shortages in funding as well as the sectoral spread of responsibilities. In international agencies, there are not many positions that focus full-time on nutrition; at a country level, similar situations lead to weak on-the-ground capacity.

The final element that is required for success is effective *use of information for awareness creation, monitoring, learning, and evaluation*. The nutrition community has suffered from insufficient translation of the evidence that does exist to decision makers at the policy level.

The Way Forward

Recently, within the international nutrition community, much discussion has focused on the lack of institutional capacity and other weaknesses. There is a real recognition that there are institutional problems to solve. There are questions about which members of the UN family will emerge to take the lead in bringing players together to define coherent messages to nonnutrition policy communities, advocating for more resources, and becoming a channel for those resources to be used.

It is unlikely that members of the nutrition community themselves are going to spontaneously solve these problems. Instead, what may be required is a high-level mandate for institutional change. Ideally, that mandate should come from leaders in the developing world through the vehicle of the G20 or other forum. Short of that, the mandate should come from some of the major funding agencies working together to lay out a set of expectations about how institutions should and could learn to allocate roles and work together.

At the same time, there need to be adequate resources or additional resources devoted to bolstering institutional capacity within the key UN agencies and others, so that there can be a response to that mandate. The institutional leadership—the key coordinating institution—needs to come from within the UN family.

Serious engagement of the private sector is an element that needs to be fostered. Significant new resources need to be brought to bear (at a minimum internationally, but ideally from affected countries themselves) to help with the processes of establishing a technical consensus, building the evidence, scaling up key programs, and monitoring their results.

THE ROLE AND CAPACITY OF FOUNDATIONS IN RESPONDING TO THE CRISES

Haddis Tadesse, M.P.A., Policy and External Relations Officer
The Bill & Melinda Gates Foundation

The Bill & Melinda Gates Foundation believes that people are trapped in poverty and ill health not because of how smart or dedicated they are, but simply because of where they were born and, consequently, whether they have access to tools, technologies, vaccines, health care, or the general environment necessary to lead a healthy and productive life. The Bill & Melinda Gates Foundation cannot be involved in every important development issue; instead, a few key questions drive its decisions about which issues to focus on: What issues affect the most people? What issues have been neglected? Where can the Foundation make the greatest impact?

The Global Development Program

The Bill & Melinda Gates Foundation gives about 50 percent of its resources to global health, about 25 percent to U.S. programs that focus mostly on education, and about 25 percent to global development, of which agricultural development is the largest program. The Global Development Program was designed with a set of key principles:

- Philanthropy plays an important but limited role. The Bill & Melinda Gates Foundation can take risks, move quickly, and catalyze change, but large-scale, sustainable change is driven by markets and governments. The Bill & Melinda Gates Foundation sees its role as complementing and strengthening governments, not competing with or replacing them.
- The focus must be on benefitting individuals. There are many ways to approach and quantify development. Individual people are the lens through which the Bill & Melinda Gates Foundation views and measures success.
- The greatest impact can be achieved by focusing on a few key, long-term issues.

More than 1 billion people live on less than $1 a day, and about 1 billion people are chronically hungry. This kind of grinding poverty often reduces life to a daily struggle for survival. Food, water, shelter, health care, and education can seem like luxuries. A majority of the people who live on less than $1 a day rely on agriculture but struggle to grow enough food to eat. Because the Bill & Melinda Gates Foundation is concerned with reducing hunger and poverty, agriculture was an obvious place to start.

The story of agricultural progress parallels the story of human progress. Over the past 200 years, nearly every part of the developed world has seen an agricultural transformation that has dramatically reduced poverty and hunger. When agriculture flourishes, so does society; where it doesn't, hunger and poverty take root. Yet agriculture in developing countries has been neglected by country governments and donors over the past several decades. That neglect has produced a lower level of food production than is necessary.

Farmers in Africa and in South Asia face starkly different circumstances than those in more prosperous regions. While of course African farms should not necessarily look like farms in the United States, they lend themselves to comparison. A typical farmer in the United States will have a large tract of land; a tractor equipped with a global positioning system (GPS), air drill, and other technologies that analyze soil; variable fertilizer applications; access to the best scientists in the world; and a global supply chain. The experience of a typical farmer in Africa is vastly different. She has a very small amount of land; no reasonable access to input or output; no roads; no extension services; no access to credit; no marketing information; and, should production fall, there is no safety net to protect her. For Africans, farming is a risky and unforgiving enterprise.

Agricultural Goal of the Gates Foundation

The goal of the Bill & Melinda Gates Foundation is simple, yet ambitious. It is to help 150 million farming families move out of poverty while significantly improving child and household nutrition. The approach begins and ends with the small farmer. Everything the Bill & Melinda Gates Foundation does is focused on him or, more likely, her. Women are at the center of these efforts, and the impact of each grant on gender is considered.

The Bill & Melinda Gates Foundation focuses on sub-Saharan Africa and South Asia simply because about 80 percent of the people who make less than $1 a day live in those regions. The results the Bill & Melinda Gates Foundation seeks include increased household incomes, child weights, and the quality and quantity of diets.

The story of one small farmer's success—of growing, harvesting, and selling—is bound in the larger story of agriculture. Success requires not only quality seeds and healthy soils, but also good information, access to markets, and supportive policies. This is why the Bill & Melinda Gates Foundation is pursuing improvements along the entire agricultural value chain. Obstacles to success in this space span sectors, and so do solutions. The Bill & Melinda Gates Foundation partners with organizations in the public and private sector in developing and developed countries.

Since its inception, the Bill & Melinda Gates Foundation has given 136 grants with a total commitment of $1.2 billion. The average size of grants is about $9.2 million. By far, the largest grantee is the Alliance for Green Revolution in Africa, which is a Nairobi-based organization led by Africans. The current chair of the alliance is former UN Secretary-General Kofi Annan; its mission is to usher a uniquely African Green Revolution on the continent. In Africa, the Bill & Melinda Gates Foundation's grants represent less than 5 percent of the total global commitment of $9 billion for agriculture and agricultural spending in Africa.

Lessons Learned

The Bill & Melinda Gates Foundation has learned that the major determinant of success is the design of programs. The Bill & Melinda Gates Foundation chooses programs that are focused on the customer—in this case, smallholder farmers, and particularly women.

Another lesson learned is that the Bill & Melinda Gates Foundation cannot merely invest in science, technology, and production. Instead, market access programs, in which farmers have the ability to market their crops, earn income, and reinvest in their own agriculture, are key to empowerment and sustainability.

Additionally, the quality of partnerships defines the success of the Bill & Melinda Gates Foundation. The organization itself does not implement any projects on the ground; in this way, partners are paramount. Public-private partnerships offer a powerful opportunity to respond to the challenges of a global problem. By combining resources, merging expertise, and sharing risks, partnerships can deliver results. For partnerships to succeed in these joint ventures, each partner needs their roles defined clearly, based on the strength that each partner brings. Results of such partnerships must be tracked over time.

For example, the cocoa partnership aims to raise incomes through improved knowledge and productivity by working with a dozen multinational corporations to double the income of 200,000 cocoa-farming families in five African countries. When a single family achieves this goal, it is a great success. When such success is experienced by entire communities, it is development. Although it is not the Bill & Melinda Gates Foundation's core competency, the foundation has initiated some short-term, emergency grants around food security during the food crisis in multiple countries.

Solutions and Investment Ideas

The following solutions and investment ideas were offered by the Bill & Melinda Gates Foundation:

- Research—If agriculture is to change for the better, the international agriculture community needs to identify and fund research with the greatest promise.
- In-country investment—Using new funding modalities, rapid and effective expansion of donor financing for agriculture is needed.
- Support local transformative efforts—On-the-ground initiatives that are tailored to local farmers need to be supported.
- Build human capacity network—A large cadre of well-trained, motivated agricultural scientists and extension workers must be created.
- Reinvest in agricultural value chain—Private companies that look beyond corporate social responsibility and find neutral business value should be identified.
- Invest in livestock—Investing in livestock is critical and contributes to nutrition, risk mitigation, and crop protection.
- Protect food security in biofuel production—Biofuel needs to be produced in a manner that protects food security.
- Expand feeding programs—Appropriate strategies for local procurement of food must be established. (The WFP is the leading agency in this effort.)
- Infrastructure—Agricultural growth and poverty reduction depend on investment in agriculture.
- Pilot large-scale programs in country—Large-scale pilot farm programs intended to stand as models should be supported.

Even with all these opportunities available, there are still great challenges and risks ahead. To succeed over the long term, agriculture needs to be both environmentally and economically sustainable for the small farmer. Climate change is a major issue that farmers are facing, and it particularly affects the poor in developing countries. The world population is projected to be 9 billion by 2050; nearly 50 percent more people will need to be fed on less arable land than currently exists today.

The Bill & Melinda Gates Foundation is optimistic about the future. Developing countries are making enormous progress. The incredible benefits from science and technology that have been witnessed over the past several years will continue. Renewed attention is being given to hunger and agriculture. Success is possible, and great partners around the world will make it happen.

THE ROLE OF FOOD COMPANIES IN RESPONDING TO THE CRISES

Derek Yach, M.B., Ch.B., M.P.H.,
Senior Vice President of Global Health Policy
PepsiCo

The international development community is off target in its effort to achieve the Millennium Development Goals (MDGs). The role of the private sector in achieving these goals is rarely discussed, but changing economic realities are leading to novel ways of doing business. There are a number of companies seeking closer integration of financial performance and social purpose (Table 6-2).

PepsiCo's CEO Indra Nooyi calls it "performance with purpose" and is proposing that the compensation of executives be linked to their financial performance and to the attainment of broader health and environmental goals (PepsiCo, 2007). The notion of "creative capitalism" is one that Bill Gates has been advancing in his effort to find the interface between financial and social gain (Gates, 2008). Mohammed Yunus has epitomized the bringing together of public-private partnerships through the Danone and Grameen Bank partnership to address the nutrition needs of poor communities (Yunus, 2008). There is now a recognized need to develop new business models that generate profits while improving population health and the environment.

A Role for Food Companies in Addressing Undernutrition

This section outlines eight actions that indicate specific roles for private food companies in addressing world hunger. These actions ideally are taken in collaboration with public and other civil society players.

Investment in Agriculture, Especially Smallholders

It often comes as a surprise that food companies actually grow a lot of their food directly or that they buy actual commodities. In fact, Lay's chips come from potatoes; a can of Pepsi depends upon cola nuts grown in Senegal and vanilla pods from the trees of Madagascar; and Quaker's oats are grown around the world in real fields. Nestlé directly employs more than 600,000 farmers around the world. The footprint of agriculture made by food companies around the world is enormous.

This creates opportunities for food companies to support smallholder farming in a number of countries, as well as to employ those who do not currently have jobs. The long-term viability of the global food supply requires government action and corporate philanthropy aimed at helping farmers access higher-quality seed, microcredit, fertilizer, and efficient irrigation systems.

TABLE 6-2 Examples of Food and Beverage Companies' Actions to Reduce Global Hunger

Company	Action	Likely Impact
General Mills	Support of local high-value corn in China	Stimulating local economy; supporting local agriculture
PepsiCo	Support of local potato farmers in Peru, China, and South Africa; citrus farmers in Indian Punjab; corn farmers in Mexico; oats farmers worldwide; and investment in pilot projects to address hunger in Nigeria, South Africa, Mexico, and India	Stimulating local economy and local agriculture; developing new products and business models to address hunger
Unilever	Commitment to World Food Programme to combat child hunger in Kenya, Indonesia, Ghana, and Colombia; creation of Annapurna iodized salt in Ghana	Reduction of child hunger; improvement of child nutrition, health, and education; increased school attendance; reduction of micronutrient deficiencies; improvement of health outcomes
PepsiCo Foundation	Investment in Save the Children for obtainment of food, deworming drugs, breast-feeding advice, and tools for proper hygiene in India and Bangladesh; invest in strengthening the World Food Programme's distribution capabilities	Reduction of child hunger; disseminating information; educating population; providing pharmaceuticals for achieving better health outcomes (overall disease reduction); disseminating information for purposes of providing better food aid to populations
Coca-Cola	Commitment to Nets for Life for distribution of malaria nets in sub-Saharan Africa	Reduction of malaria and malaria-related diseases
Nestlé	Development of Popularly Priced Product portfolio to offer food and beverage products high in nutrient quality for consumers in poverty	Reduction of hunger; improvement of health outcomes
Heinz	Early supporter of Sprinkles' satchel of micronutrients and Britannia's iron-fortified biscuits in India	Reduction of micronutrient deficiencies; improvement of health outcomes
Danone	Creation of innovative social business partnership with Grameen Bank in Bangladesh for distribution of nutrients to undernourished populations	Reduction of micronutrient deficiencies; improvement of health outcomes; support of local agriculture

*Expanded Use of Corporations' Core Capabilities
in Distribution and Quality Control*

The private sector's efficient distribution systems—particularly those of companies like Coke, Pepsi, Nestlé, Unilever, and Procter & Gamble that distribute small consumer goods—could be leveraged to address hunger and malnutrition. Global food companies can leverage distribution expertise to help NGOs and the WFP develop successful strategies for reaching the hungry. For example, an evolving partnership between the PepsiCo Foundation and the WFP places PepsiCo retirees with distribution expertise in positions within the WFP in order to increase distribution capability in other countries. Expanding such systems could build upon food companies' expertise in food quality, forecasting, and marketing (Yach, 2008).

*Sustained and Greater Support for Fortification of Staples and
Commonly Consumed Nutritious Foods and Beverages*

Micronutrient deficiencies affect 2 billion people worldwide and contribute to birth defects, stunted development, and disease. Many food companies have taken their role in food fortification very seriously, especially when their products are widely consumed in the developing world. The Global Alliance for Improved Nutrition, through its corporate partnership programs, assists food companies in using their current product lines to meet nutrition criteria. While there is no intent for classic Pepsi to become a superfortified product, many products within the increasingly healthier side of PepsiCo's portfolio could and should be fortified, especially in countries with major micronutrient needs. To further such processes, food companies need partners in many countries because fortification programs are not simple to implement, assess, or monitor.

*Innovation and Expansion of the Portfolio of Foods Currently
Available for Complementary Feeding in Settings of Acute and
Chronic Undernutrition and Support for Breast-Feeding*

Food companies could be far more forceful about their support for breast-feeding. Unfortunately, the centrality of breast-feeding tends to slip off the agenda when other products are being promoted.

Ready-to-eat therapeutic foods are another area where food companies can and should play a role. PepsiCo is working with Valid International to support the development of ready-to-use therapeutic foods in Nigeria. There is a huge demand for food and food products that is currently not being met in crisis areas; this is a "window of opportunity" for food companies to play a role in preventing the worst effects of food crises like the current one (Sheeran, 2008).

Cocreation of New and Innovative Social Business Models to Help Combat the Global Burden of Undernutrition

A satisfactory business model for many companies may actually be one that simply breaks even. Muhammed Yunus's idea for "social business" is a nonloss, nondividend company in which profits are reinvested to expand reach and improve the product. In partnership with Danone and Grameen Bank, Yunus succeeded in delivering nutrients to undernourished individuals in Bangladesh. Other companies may demand a marginal or very small profit margin or a profit margin that will extend over decades.

If food companies find ways to develop these social business models, a large amount of resources would be liberated within companies (although this would require that Wall Street tolerate companies doing this). Given the new environment since the economic crisis, such models are more likely to get a higher degree of support than they would have in the past.

Investment in the Development of Nutrition Science Capacity, Especially in Developing Nations

Food companies' investment in the development of nutrition science capacity is not for philanthropic reasons, but rather for self-interest. Top nutrition and medical journals focus overwhelmingly on obesity while ignoring India, China, and Africa, which make up 40 percent of the total world population. In such developing countries, neither food companies nor public health agencies can find the nutrition scientists needed for their work. In this way, it is in food companies' best interests to invest in building the science base in nutrition in developing countries. In particular, PepsiCo supports the International Union of Nutritional Sciences, the International Nutrition Foundation, and the Tufts School of Nutrition initiative to identify ways of building a global plan to enhance capacity for nutrition science in low- and middle-income countries.

Innovative Product Reformulation Aimed at Developing Low-Cost Nutritious Foods for All Markets

Food companies are uniquely more capable than the public sector at making foods tasty and fun. At the same time, tasty and fun does not have to mean not healthy and nonnutritious. In fact, the key to increasing the quality of nutritious food is to focus on taste, recognizing that of course the tastiest and healthiest foods will always remain fresh fruit and vegetables, but fresh fruits and vegetables simply cannot solve all the massive undernutrition problems the world is facing.

Committed Advocacy by Multinational Food and Beverage Corporations for Nutrition-Friendly Trade Policies

Food companies can and should support significant changes to trade policies that encourage subsidies that make it easier to grow and eat nutritious foods. A change in policies around subsidies could have a big impact on agricultural production in developing countries as well as incentivizing food companies' expansion into areas that address undernutrition and other health issues such as cardiovascular disease.

THE ADVOCACY ROLE OF CIVIL SOCIETY ORGANIZATIONS IN RESPONDING TO THE ECONOMIC AND FOOD PRICE CRISES

Asma Lateef, B.A., M.A., Director
Bread for the World Institute

Overview

Bread for the World Institute is the research and education affiliate of Bread for the World, a collective Christian voice urging decision makers to end hunger at home and abroad. Bread for the World is a grassroots advocacy organization with a network of 60,000 members, churches, and denominations in the United States. Bread for the World Institute provides the research and policy analysis that informs Bread for the World's legislative advocacy. Bread for the World mobilizes its network to generate personalized letters, e-mails, and phone calls to members of Congress on particular legislation around hunger and poverty.

Civil society organizations (CSOs) like Bread for the World Institute need to capitalize on the current opportunity (the best in decades) to get nutrition on the global agenda. The food price crisis and the economic crisis have shone the spotlight on the vulnerability that exists in the developing world. The current crises can serve as a clarion call for advocacy organizations to help initiate a response to not only the crisis but also the underlying chronic conditions that exist. CSOs can play a distinct advocacy role in shaping the global response to the global economic crisis, particularly the focus on food security, agriculture, and nutrition. Opportunities abound for CSOs to engage governments and international institutions on what is happening on the ground, policy recommendations, and implementation.

Civil Society's Advocacy Role

There is a role that advocacy organizations can play to engage in the policy debate within the CSO community. CSOs can educate and rally by drawing attention to and defining the issues, building consensus among policy advocates, and working together to shape and advance policy solutions and recommendations. Additionally, CSOs have a role in the policy debate with the administration and Congress. CSOs can persuade through elevating the importance and urging leadership on an issue; they can engage in the debates that are raging within or between Congress and the administration; and they can play a brokering role by carrying information and messages between different players, identifying common objectives, and adding pressure when needed.

CSOs have a further role to play in engaging and mobilizing grassroots constituents. For example, the media frenzy around the current food crisis was very helpful in the United States as it galvanized a series of very high-profile initiatives. The *Lancet* series came just as there was growing awareness of the food price crisis, but the ensuing food riots, particularly from a national security angle, grabbed the attention of policy makers.

The Center for Strategic and International Studies sponsored a Task Force on the Global Food Crisis chaired by Senators Lugar and Casey. A CSO coalition came together to develop the *Roadmap to End Global Hunger*. The Chicago Council on Global Affairs did a report on global agricultural development. This series of high-profile events created momentum. CSOs can play a role by showing the connections between these issues and how specific policy actions can help. CSOs can build relationships with congressional offices and staff to show there is a long-term commitment to the issues.

The initial policy advocacy toward the global food crisis was focused on the short-term response—boosting immediate agricultural production, safety nets, and cash and food assistance programs. As the Food and Agriculture Organization of the UN (FAO) and others updated alarming food insecurity and hunger data, advocacy efforts expanded their focus to include addressing malnutrition. There is a growing recognition that an important part of a more comprehensive food security strategy must focus attention on addressing and preventing malnutrition, particularly during the first 2 years of life.

Recent and Future Advocacy Opportunities

Advocacy organizations in the United States focus on the U.S. administration and government's response. High-level global meetings are an opportunity for CSOs to engage government on their preparations and plans for these international meetings.

At the global level there have been a number of events:

- June 2008 Summit on Food Security in Rome,
- July 2008 UN High-Level Task Force and the Comprehensive Framework for Action,
- March 2009 World Bank Roundtable,
- April 2009 G20 Summit in London,

- July 2009 G8 Summit in L'Aquila, Italy,
- September 2009 G20 Summit in Pittsburgh, Pennsylvania, and
- Annual UN General Assembly meeting.

At the national level in the United States, Congress recently passed an emergency supplemental bill that included funding for agriculture development and for such short-term emergency responses as getting seed and credit out to farmers. It also included funding for food aid and a pilot project for local and regional purchase of food aid. The United States has also recently restated its commitment to achieve the MDGs, which is another tremendous step forward. Senators Lugar and Casey introduced a bill called the Global Food Security Act that was focused on agriculture, influenced by the conversations around the Center for Strategic and International Studies report. The House version of that bill, which has just been introduced by Representative McCollum, includes maternal and child undernutrition as part of a comprehensive approach.

At the national level are these initiatives:

- Emergency supplements,
- Meetings with the Obama transition team,
- Lugar-Casey/McCollum bill,
- McGovern/Emerson Roadmap legislation,
- Appropriations,
- Obama administration's Food Security Initiative,
- Foreign aid reform, and
- Quadrennial diplomacy and development review.

All of these initiatives are occurring in the context of a larger "rethink" within the administration and Congress about the way the United States provides foreign aid with a focus on country ownership, outcome-oriented development programs, capacity building, and policy coherence. Representative Howard Berman, the chairman of the House Foreign Affairs Committee, has introduced a bill that calls for the administration to develop a national strategy for global development. At this critical moment in time, there is an urgency for CSOs to engage the public and the government on nutrition issues.

THE ROLE AND CAPACITY OF CIVIL SOCIETY IN RESPONDING TO THE CRISES

Tom Arnold, M.S., Chief Executive Officer
Concern Worldwide

There is a sense of opportunity amidst the current global food crisis—a hopefulness that the hunger agenda will become a political focus. This opportunity is a direct response to the food price crisis that brought the issue of food security to a level of "political influenza," more so than at any time since the early 1970s.

Money is not the key to solving the problem; the key is policy change. What is the political economy of change in regard to these issues? Where are the levers of power and influence that will wake people from a current state of relative comfort with 1 billion hungry people in the world? How can this food and nutrition agenda achieve significant change? Change cannot be brought about by governments acting on their own. It must be brought about by new forms of partnerships between civil society, the private sector, and foundations.

Innovation on the Ground

The larger international NGOs can have significant impact through innovation on the ground in the countries where they work. For example, Concern Worldwide and Valid International were the pioneers in developing the "community-based management of severe acute malnutrition" approach. In 2000, Concern Worldwide initiated this approach in three countries. Through the processes of group program design, proper data collection, and sharing that information within the nutrition and wider community, policy change at the international level was achieved. Other work in the area of safety nets and conservation farming are equally promising. Such successful initiatives need to be encouraged and expanded.

New Models of Partnership

Concern Worldwide has significant and important partnerships with the International Food Policy Research Institute (IFPRI) and the Cary Group—the largest Irish international food company. These partnerships look at action research on the ground, nutrition status, livelihoods, and HIV/AIDS data. Such partnerships are becoming common between NGOs and business.

Other opportunities exist for partnerships between businesses and higher education institutions. Such partnerships should be seen as learning communities, where the results that are generated from well-designed programs can feed into a learning loop for the benefit of all.

Advocacy

Civil society must play a crucial role in advocacy at the national and international levels if the hunger agenda is to change. The UN High-Level Task Force has created a virtual network that connects work at the UN level with all of the major NGOs working in the field. At the European level, the European Food Security Group brings together all of the major European NGOs and engages directly with the European Commission and the European Parliament. In the United States, the *Roadmap to End Global Hunger* is a similar, very positive initiative, and should be further built upon. The Global Hunger Index is a document produced by IFPRI that ranks every country in regard to hunger on an annual basis; this document is becoming an important advocacy tool (Grebmer et al., 2009).

Leadership in Ending Hunger

Perhaps most important of all is the concept of leadership—at national and international levels—and how it can be the stimulus for change. If the hunger problem is to be solved, the critical locus of leadership must be at the national level for food-insecure countries. Unless there is policy change at the national levels, all of the good intentions at the international level will not lead to change. That crucial connection between the global and the national agendas must not be forgotten.

The comprehensive framework for action, which arose from the Rome summit in 2000, provides a sensible framework for policy change going forward—a set of short- and longer-term actions that need to take place. These actions must be supported by resources from donors, such as the $20 billion commitment from the G8. There is a nonfinancial dimension to the $20 billion commitment as well. Such an international political commitment has additional power to change policy at the national level.

The United States has a critical role to play, and there seems to be a shift in current U.S. policy from food aid to farming. Undergoing such a transition, however, will trigger challenges in dealing with the "iron triangle"—the farmers, the shippers, and the aid agencies—who will not want such change.

The European Union (EU) is important as well; the fact that the EU is beginning to think about the importance of nutrition is significant and important. In particular, Ireland has an important role to play. The Irish Hunger Task Force produced a report last year that is widely recognized as being a quality product focusing on three areas: (1) promoting the productivity of small-scale African agriculture, (2) putting a very clear focus on nutrition, and (3) Ireland playing a leadership role in advocating for change at the global level.

MITIGATING THE NUTRITIONAL IMPACT OF THE GLOBAL FOOD SECURITY CRISIS: THE ROLE AND CAPACITY OF UN AGENCIES IN RESPONSE TO THE CRISIS

David Nabarro, M.D., Coordinator
UN System High-Level Task Force on the Global Food Security Crisis

Context

From the perspectives of the world's 2 billion poorest people, food and nutrition systems were in crisis long before food prices shot up in 2007. Economic contraction made the situation worse. Because of high levels of malnutrition and growing hunger, increasing numbers of people in resource-poor settings are unable to realize their right to food—a right to access the food they need, when they need it. Food production, processing, and distribution systems are not serving the interests of small-scale producers, as evidenced by the fact that there are great differences between the prices paid to producers and the prices paid by consumers. Volatility in food prices and variations in supplies are impeding global progress toward the MDGs. These are all very complex issues that are symptoms of seriously dysfunctional systems.

High-Level Task Force on the Global Food Crisis

The structure of the High-Level Task Force (HLTF) is such that it is a time-limited UN system entity established at the end of April 2008 by the UN Chief Executive Board. The HLTF includes the WFP, FAO, International Fund for Agricultural Development, The World Bank, International Monetary Fund, UNICEF, UN Development Progamme, World Trade Organization, UN Conference on Trade and Development, the International Labour Organization, and 14 other UN entities. The UN Secretary-General, Ban Ki-moon, chairs the HLTF, and Jacques Diouf, the FAO director general, is vice chair. The HLTF has one coordinator and a small secretariat.

The mandate of the HLTF is that it is not to be a new organization; it was developed to avoid creating new bureaucratic and intergovernmental layers. It is simply a mechanism to try to make the UN system work in a more integrated and coordinated way on policy and program issues. The goals of the HLTF are to enable access to food in the *short-term* through improved nutrition, social protection, and food systems (taking account of life-cycle vulnerabilities); and in the *longer-term* to improve availability of food with an emphasis on small-scale agriculture and making markets and systems for trading foods function in the interests of poor people. The unifying principles of the HLTF are a commitment to MDG 1; bridging short- and long-term responses to try to end the divide between humanitarian development and trade approaches to food issues; and an attempt to have an integrated response.

The HLTF is a construct of the agencies under the chairmanship of the UN Secretary-General, who has a strong role as the chair; it is not a member state construct. The main outcome of the HLTF aims to have UN agencies work well together at the country level. It has been recognized that there are turf battles and inconsistencies between different agencies on the ground that lead to a lack of credibility in the eyes of national governments and donors. The Committee on Food Security is a completely different entity. Its aim is the governance of food security in member states through relationships with civil society and producer organizations in the private sector. The HLTF is a subset of, and subject to, the Committee on Food Security. The HLTF is on the implementation and agency side, not the governance of member states side.

The HLTF developed the Comprehensive Framework for Action as a means for organizing collective actions in pursuit of immediate and longer-term outcomes to be taken forward by different stakeholders working in partnership (not limited to the HLTF membership) under the leadership of national authorities.

The Comprehensive Framework for Action has two objectives. The first is to improve access to food and nutrition support and to take immediate steps to increase food availability. The desired outcome of this objective is to meet the immediate needs of vulnerable populations. This will be accomplished by:

- Offering emergency food assistance and nutrition interventions, and enhancing safety nets and making them more accessible;
- Boosting smallholder farmer food production;
- Adjusting trade and tax policies; and
- Managing macroeconomic implications.

The second objective of the Comprehensive Framework for Action is to strengthen food and nutrition security in the longer run by addressing the underlying factors driving the food crisis. The desired outcome of this objective is to increase longer-term resilience and global food and nutrition security. This will be accomplished by:

- Expanding social protection systems;
- Sustaining the growth of smallholder farmer food production;
- Improving international food markets; and
- Developing an international consensus on biofuel production.

The HLTF focuses on four lines of work. The first is to support realization of the Comprehensive Framework for Action outcomes in countries. The HLTF will respond both to country needs and requests from national authorities. At the national level in the majority of countries, there has not been a comprehensive approach to food insecurity undertaken. Instead, there has been compartmentalization of the response. Second, the HLTF has been advocating for the resources needed for urgent action as well as long-term investment based on the right to food, the need for adequate nutrition, and reduction in vulnerability. Third, the HLTF is trying to inspire a broader engagement in food issues so that the hundreds of stakeholders involved in food security issues in any one country can come together in a concerted movement for food security. Fourth, the HLTF would like to ensure accountability through assessing achievements, reviewing progress, demonstrating results, adjusting activities that are suboptimal, and presenting reports.

High-Level Meetings on Food Security

The High-Level Meeting on Food Security for All in Madrid in January 2009 was a very important initiative. It challenged a number of different actors to find ways to stimulate more resources and achieve more open consultations and partnerships for food security. One billion Euros were pledged at this meeting, but more importantly, there was a clear consensus that the way forward was the twin approach of the Comprehensive Framework for Action (immediate and long-term), with addition of FAO *Right to Food Guidelines* as a basis for analysis, action, and accountability. The need for a mechanism to properly coordinate financing as well as a need for extensive and open consultations for establishing partnerships was also expressed.

At the G8 Initiative on Food Security in L'Aquila in July 2009, there was a focus on comprehensive (both short- and long-term) food production, access, and use. Additionally, there was a clear recognition of the role of agriculture, not only for food production, but also as a means for community resilience and as a base for economic growth and development.

In L'Aquila, there was real attention paid to vulnerabilities, particularly of young children and women, and to the commitment to support national efforts and regional initiatives to improve food security as well as to ensure a coordinated international response. There was emphasis on engaging producer organizations and civil society on the importance of the private sector and on producers having greater access to the value chain.

A commitment of $20 billion in food and agricultural assistance to those suffering the effects of the food crisis was also made at the G8 initiative. President Obama managed to increase the sum from $15 billion to $20 billion because of the intensity with which all the leaders (not only the G8—26 countries were represented) supported the need for a new initiative on food security. The $20 billion commitment is simply a statement of intention, and no one knows precisely how much of this is "new" money. But it is important to give adequate attention to those at risk of malnutrition and affected by chronic hunger within this new thinking.

Challenges and Solutions

Outstanding and challenging policy issues remain. They include ensuring the generation of knowledge and its application to:

- Enable populations to realize their right to food,
- Improve humanitarian access to food,
- Incorporate nutrition-related analysis and action in response,
- Examine and pursue options for predicting future crises and mitigating them (e.g., through price stabilization),

- Pursue food-trading systems that benefit poor people,
- Promote best practices for conservation agriculture,
- Respond to the needs of marginal populations (especially pastoralists), and
- Promote the development of biofuels that do not compete with food crops.

As the world moves into a prolonged period of economic downturn, national authorities and international organizations have concerns that the needed private and public investments in agriculture, social protection, and a fair trade system will not be immediately forthcoming. The structural problems in food and agriculture may persist, as will the risk of extreme volatility in prices.

What can be improved upon is an increased recognition of the likely magnitude and duration of the food security crisis in different countries. The international food security community must be able to adapt its response to changing circumstances and emerging threats to food security such as volatility in food, energy, and financial markets; climate change; and water scarcity. The international community will need to mobilize and use the necessary resources while sustaining political commitment.

There is a need to improve funding of food assistance, smallholder production, and social protection. Options to do so include coordinating funds provided through existing funding channels (e.g., the World Bank, International Fund for Agricultural Development, WFP, FAO, regional banks, bilateral donors, and foundations) so that they respond to country needs in an integrated way, or establishing novel funding mechanisms that pool donor resources and react in a predictable way to country needs.

Within that context, as international organizations align themselves, they are functioning within a much broader, more substantive, and novel form of partnership, built around a joint understanding of the issues, an understanding that there are opportunities to act, and an eagerness to achieve that goes beyond static rigid structures and move toward empowering and open partnerships, almost social movements. In this new paradigm, institutions must take joint responsibility for ensuring that results carry through. This is a new way of working—separately, but together; strategically *and* opportunistically.

THE ROLE AND CAPACITY OF UNICEF IN RESPONDING TO THE CRISES

Werner Schultink, M.D., Chief of Child Development and Nutrition
UNICEF

Background

The United Nations Children's Fund (UNICEF) has a presence on the ground in more than 100 countries. In most of these countries, UNICEF also has staff especially focused on nutrition and public health. However, looking back at the end of 2006 or 2007, much of UNICEF's nutrition work was focused on the prevention of mortality and, in this way, had less of an impact on MDG 1. In 2007, just before the *Lancet* publication was released, UNICEF underwent a review. The information in the *Lancet* was used, and a significant number of experts reviewed UNICEF's nutrition programming.

As a consequence of that review, nutrition programming was modified to place more emphasis on nutrition interventions during pregnancy and in adolescence. UNICEF attempted to significantly enhance its work in infant feeding, including breast-feeding and complementary feeding.

In the UN system, UNICEF plays the lead role for interventions that mitigate the impact of emergencies, acute as well as prolonged. Taking the lead in these efforts has considerably strengthened the collaboration that UNICEF has with its sister agencies, especially the WFP, WHO, and FAO. For example, the REACH initiative involved four UN agencies and a significant number of NGOs who agreed upon a core set of interventions and approaches that are being piloted in Laos and Mauritania.

UNICEF Reaction to Food Price Increases

In July 2008 when food prices escalated, UNICEF was experiencing extremely limited progress in the reduction of malnutrition. Between 1995 and 2005, UNICEF estimated that the reduction in stunting was at best a reduction from 33 percent to 30 percent, and the wasting rate hovered around 10 to 11 percent during the entire time period.

When the food crisis occurred, UNICEF participated in the initiation of the High-Level Task Force of the Secretary-General, which proposed two main pillars of intervention—one to provide immediate relief and the other to look at the required long-term action to improve the sustainability and the availability of food.

UNICEF focused on immediate relief and the provision of nutrition security for vulnerable groups. The phrase *nutrition security* was chosen deliberately, because UNICEF feels that household food security is very different from nutrition security, especially when considering the consequences of undernutrition for young children. If action is not taken at a very early age, the consequences of undernutrition can lead to lifelong negative impacts, and, ultimately, an entire lost generation.

UNICEF applied $50 to $60 million of its own resources to stimulate a package of immediate relief with a focus on nutrition security for vulnerable groups. Throughout this process, UNICEF had regular contact with the WFP and WHO on the types of interventions supported, the role of other agencies, and which countries would receive UNICEF's focus. This is an indication of the strong collaboration between UN agencies during the crisis.

Forty-two countries were selected, based on high under-five mortality rates, high malnutrition rates, and high HIV rates. UNICEF realized that the high food prices may have an impact on households affected by HIV, who likely already manage on a very meager income and are net buyers of food.

Impact of UNICEF Response to Food Price Crisis

UNICEF's funds were distributed quite rapidly. Almost all of the selected countries received funds (over and above their already available budgets) by the end of July, and 70–80 percent of these funds were actually used. About 80 percent of the funds were used for the expansion of nutrition and health programs; 28 countries worked on evidence building and the enhancing of data availability. For example, 9 countries in West Africa conducted smart surveys, seven are in the pipeline, and a number of other studies and data collection efforts were made, including the strengthening of surveillance in Zimbabwe.

Another noted impact was a rapid increase in the treatment of children suffering from severe and acute malnutrition. In 2007, UNICEF distributed about 4 million kilograms of ready-to-use therapeutic foods, and in 2008 that number jumped to 10 million kilograms. With this increase, an estimated 750,000 to 1 million children suffering from severe and acute malnutrition were treated.

The use of multiple micronutrients was also expanded. For example, in Afghanistan, Cambodia, Indonesia, and Laos, large groups of people were reached (with a focus on children 6 to 24 months of age) with micronutrient supplements to improve the quality of their complementary food. A significant number of countries, such as Angola, the Congo, Ethiopia, Mali, Malawi, and Kenya, further improved their infant feeding programs.

Also accompanying this outreach was an increase in the capacity of national institutes and NGOs to carry out this work. To expand outreach, basic community workers and health workers in the field must be enabled to implement programs and educate mothers. Much capacity strengthening occurred in 25 countries, including Benin, Cambodia, Liberia, Mozambique, and Zambia. For example, prior to the crisis in Mozambique, the provision of immunizations, impregnated bed nets, vitamin A supplements, and deworming medication was effectively accomplished only once a year. In 2008, as a consequence of the food crisis and with UNICEF support, Mozambique modified its programs to be implemented twice a year, leading to an enormous increase in coverage with essential interventions. Finally, UNICEF also supported a number of social safety net programs. For example, cash support was provided in Madagascar, Ethiopia, and Malawi.

Without knowing to what extent these interventions will translate into lasting changes in wasting or stunting rates or in other indicators, it is safe to assume that at least a number of indicators will have shifted. UNICEF has reached more children with treatment of severe and acute malnutrition and, in this way, prevented child deaths. A substantial number of children were reached with vitamin A supplements, and the first steps were taken to improve infant feeding in a number of countries. The focus on "nutrition security" is of crucial importance when considering the lives of children and the consequences they will face in later life.

THE ROLE AND CAPACITY OF THE WFP IN RESPONDING TO THE CRISES

Martin Bloem, M.D., Ph.D., Chief for Nutrition and HIV/AIDS Policy
World Food Programme

Evolution of the WFP's Focus on Nutrition

Over the past 2 to 3 years, the WFP has undergone significant changes. Owing to the new leadership of Josette Sheeran in combination with external circumstances, the agency developed a new strategic plan. The transition has been quite dramatic as the WFP has moved from a food agency to a food assistance agency.

Twenty years ago, the WFP provided direct food assistance and food transfers, mainly from Europe, the United States, and Canada to lower-income regions. In contrast, today almost 70 percent of the WFP's budget is in cash, so the WFP is able to buy food locally. Local procurement is different from shipping food internationally; it is cost-effective, but has required new strategies in food assistance.

Another major deviation from the past at the WFP is a move away from the focus on acute hunger; today, the WFP focuses on acute *and* chronic hunger. Suddenly, programs must focus on stunting and other chronic problems. This transition occurred through much consultation with the WFP's donors because it has many consequences in how such problems are tackled.

In the context of the new strategic plan, the focus on nutrition is taken very seriously. Josette Sheeran declared nutrition as extremely important for the WFP because it is necessary for an agency that delivers food assistance to 80 to 100 million people every year to have a positive impact on nutrition. The WFP no longer focuses merely on delivering food, but also on "what is in the food basket." The connection between *access to food* and *nutrition* has not always been assumed, so this too is a new way of thinking. The WFP designed a "nutrition improvement strategy" that is based on the *Lancet* series and designates an important role for the private sector.

The WFP's Role in Responding to the Food Crises

In an effort to align the WFP's strategy for responding to the current food price and economic crises with other UN agencies, an issue brief was written. This document described how to improve nutritional status under the current circumstances, focusing intently on the importance of micronutrients.

Early in 2009, the Gandolfo meeting was held in Rome to discuss the potential implications of the economic crisis, climate change, and high food prices on nutrition. A consensus from this meeting was published in the *Journal of Nutrition*. Another outcome of the Gandolfo meeting was an issue brief with the following recommendations. First, more funding is needed in the field of nutrition to implement all of the new, proven strategies. A second recommendation is to focus on target groups, in particular minus 9- to 24-month-old children and women. Historically, the WFP did not focus on particular groups, so this is a fundamental change that aligns the WFP with the broader nutrition community. The third recommendation is that inadequate diets need to be supplemented with micronutrient powders or lipid-based products. The fourth recommendation is to enable access to nutritious foods. In the past, UN agencies only focused on the known market-based production channels. Now, it is broadly understood that it is also important to focus on the market base, and to do this in two different ways—through buying locally and partnering with the private sector.

One WFP program, the P4P program, works with small farmers so they can buy locally. Local products, however, may not be capable of improving nutritional status, so a value chain is needed. The community's food processing needs improvement through the help of local industries (supported by larger companies).

The WFP has many partnerships with the private sector. One noteworthy partnership is with DSM Nutritional Products, a Dutch vitamins and minerals company. The WFP's partnership with DSM began in 2007 with an interest in novel food products, funding, and enormous technical support that led to great progress in a short period of time. The WFP has benefitted greatly from this partnership and has pushed DSM to work with local customers in WFP recipient countries. For example, the WFP uses micronutrient powders that are not available in local markets, so must be imported. If DSM were to sell to local markets, the WFP could purchase micronutrient powders in the local market. An added benefit of a local supply of such products is that they don't require extensive social marketing because local populations become aware of the products on their own. Today the private sector plays a much larger role in the WFP's programs than merely funding. Last year, the WFP's budget was around $5.6 billion, and the private-sector contribution was about $100 million in total.

Partnerships at the local level are extremely important. The WFP's new strategic plan involves partnerships with local governments, local NGOs, international NGOs, local private-sector organizations, and international private-sector organizations. While partnerships come with their own challenges, this new model is a promising solution that is only possible when both parties have common local data, analysis, and goals.

THE ROLE AND CAPACITY OF FAO IN RESPONDING TO THE CRISES

Hafez Ghanem, Ph.D., Assistant Director-General,
Economic and Social Development Department
Food and Agriculture Organization

Reforming Global Governance for Food Security and Nutrition

The last major food crisis to face the world was in the mid-1970s. At that time, there was a call for strengthened global governance of food security, which led to the 1974 World Food Conference, and the establishment of the World Food Council and the Committee on World Food Security (CFS). The CFS was established to serve as a forum in the United Nations system to examine major problems and issues affecting the world food situation and to review and follow-up on policies concerning world food security. The Committee was useful when it began and played a large role in supporting major initiatives in support of world food security in the mid-1970s, including a call for funding for the Green Revolution and investments in South Asia, which led to a reduction in hunger.

But the world has changed. In 2009, after some progress in hunger reduction until the mid-1990s, there is again a large increase in the number of hungry people in the world, coupled with decreased investment in agriculture. The current trading system in agriculture is neither fair nor efficient. Biofuels provide opportunities for some farmers, but also present great risks for food security. Climate change is also playing a significant role in agriculture and food security, where agriculture can both be a great *contributor* to carbon emissions as well as be greatly *affected* by climate change. More recently, the impact of high and volatile food prices and the financial and economic crises have exacerbated the already high levels of chronic hunger and malnutrition in the world.

The importance of these factors and their implications for world food security all highlight the need for a renewed mechanism of governance to address fundamental weaknesses in the mechanisms governing global food security. There is general agreement that the CFS's performance has been inadequate, which leaves a vacuum in the global system of governance. The CFS should be renewed to focus more on key issues affecting global food security and include broader participation of key partners in the design and implementation of a food security policy agenda at global, regional, and national levels.

In order to improve upon the global governance structures, the members of FAO have begun the process of reforming the Committee to make it more focused and useful in addressing world food security issues. The process is ongoing. The overall goals are to redefine the role of the Committee on World Food Security so that it can be more action-oriented and to expand Committee membership beyond FAO's member states, to include broader participation from other key stakeholders, such as representatives from the NGO/CSO private sector and other UN agencies.

FAO Reform

FAO itself is also undergoing a major reform. An extensive external evaluation of FAO—the first one of its kind for a major UN agency—concluded that if FAO did not exist, it would have to be reinvented. However, FAO as it currently exists needs major changes.

The FAO reform involves introducing new or revised structures and mechanisms, including for management of the organization, new personnel systems, a new organizational culture, and new and improved governance of FAO itself. This reform is an important part of FAO's response to the current economic and financial crisis because unless the institution is modernized and able to seize opportunities, it will not properly address those most in need.

For decades, FAO has been arguing and advocating for the importance of investing in food security and nutrition; for decades it has been ignored. Suddenly, politicians are paying more attention and are now demanding that FAO and the broader international food, nutrition, and agriculture communities respond and rise to meet these new challenges. FAO must seize this opportunity to become a more effective and efficient organization in order to help address the needs of the billion undernourished people in the world.

Initiative to Support Farmers During the Food Price Crisis

A specific FAO initiative in response to the food crisis is to help farmers in developing countries increase their productivity in the short run through supply of seeds and fertilizers. This initiative started with seed money from FAO and the Spanish government, but now obtains major support from the European Commission and covers 88 countries.

As one would expect (especially economists), when food prices go up, supply responds and production increases. This was evident in response to the higher food prices of 2008, where the supply of cereals for the world increased by more than 10 percent. Interestingly, that increase in supply came almost exclusively from developed countries. Developing countries' output increased by only 1.5 percent—all from India, China, and Brazil. Poor farmers in developing countries were unable to buy fertilizers or seed, and had insufficient access to markets. For them, the high prices were not an opportunity they could seize.

FAO's initiative supports farmers in the developing world in order to allow them to seize the opportunity offered by high commodity prices. It is a short-term initiative, because it does not deal with the fundamental structural problems in the agricultural system, but it is important to allow these poor farmers to respond.

THE ROLE AND CAPACITY OF WHO IN RESPONDING TO THE CRISES

Francesco Branca, M.D., Ph.D., Director,
Department of Nutrition for Health and Development
World Health Organization

WHO Response to the Food Crisis

WHO responded to the global food crisis by pursuing the following tasks:

- Monitor the health and nutritional status of member states' populations.
- Support countries in scaling up nutrition action, such as managing severe malnutrition, promoting breast-feeding and complementary feeding practices, improving access to specific micronutrient supplements, delivering primary health care services, and promoting food hygiene.
- Support countries in strengthening and implementing integrated national nutrition policies.
- Develop and scale up social protection actions related to health and nutrition, such as working with member states to promote free or low-cost health services.
- Support member states in assessing and addressing the health and nutritional effects of food insecurity, including building capacity and training national counterparts and WHO teams, as well as designing plans and programs that can mobilize resources through the Comprehensive Framework for Action, UN Central Emergency Response Fund, and other international mechanisms.

Strategic Focus

WHO has more than 100 offices, but not many are dedicated to nutrition exclusively. About 20 percent of staff has some focus on food and nutrition. In reaction to the crisis, then, WHO was forced to reflect on its role and strategy in order to optimize its resources for nutrition. WHO produced a strategic paper suggesting that the organization focus on the following areas:

- Development and operationalization of integrated food and nutrition policies.
- Intelligence of needs and response.
- Development of evidence-based program guidance.
- Country-level advocacy and technical assistance.

WHO is seen as a "convener" within the UN family, so these initiatives were additionally proposed to WHO's sister agencies.

In the area of food and nutrition policies, WHO has been analyzing member countries' readiness to accelerate action in nutrition. The objective of such analysis is to assess existing gaps and constraints and to identify opportunities to integrate and expand nutrition-related actions in countries. An action plan is developed with recommendations to guide harmonized action for improved nutrition. Additionally, a baseline of current nutrition status and nutrition action in the 36 countries with the highest burden is established.

As a normative agency, WHO must provide technical guidance to member states. Offering guidance in collecting nutrition information and establishing national nutrition surveillance systems is an important normative role WHO plays. A new global database, the Nutrition Landscape Information System, will be launched very soon. This system provides an interface of country profiles with all of the country data collected through WHO's nutrition databases, as well as some of the data from sister agencies on policy implementation and other food security indicators.

Partnerships

WHO believes in the importance of working in partnerships and is engaged in a number of partnerships, including the UN Standing Committee on Nutrition and the REACH initiative. WHO also works at the regional level to create partnerships.

A new initiative, the Pan American Alliance for Nutrition and Development, was created to propose and implement comprehensive, intersectoral coordinated programs that are sustained over time, operate within the framework of human rights, and take a gender-sensitive and intercultural approach that contributes to reducing malnutrition and accelerating the attainment of the MDGs. This new alliance will look at effective interventions and advocate in high-burden countries for greater action in nutrition (WHO, 2009).

DISCUSSION

This discussion section encompasses the question-and-answer sessions that followed the presentations summarized in this chapter. Workshop participants' questions and comments have been consolidated under general headings.

Leadership

It was suggested by some workshop participants that UN agencies should take the lead in nutrition issues. At the same time, the view was expressed that nutrition departments in UN agencies rank low in the organizations; as a result, individuals with nutrition expertise are not in senior leadership roles.

Some leadership mechanisms were explored, such as the H8 mechanism, which is used in health. While perhaps not particularly effective, it does guarantee that the heads of agencies and important organizations talk about a topic on a regular basis. Without such a mechanism, when the heads of UN agencies come together, the chance that they talk for a significant period of time on nutrition is quite minimal.

The Standing Committee on Nutrition (SCN) was discussed. At the moment, the SCN is not officially related to any higher-level UN reporting mechanism—not to the heads of agencies, not to the UN Economic and Social Council, not to any other chamber, which makes it a voluntary mechanism that could be more effective if structured differently.

Nutrition Interventions Can Deliver

It is helpful for programs to convince funders that programs can change nutritional status. For example, vitamin A and salt iodization programs have been successful in delivering change in nutritional status at a relatively low price. Recently, many countries in sub-Saharan Africa have demonstrated substantial increases in breast-feeding rates. If these success stories are documented by countries, a sense of trust that nutrition programs can "deliver" is built with donors. In this way, it is the nutrition community's responsibility to convince people that nutrition is important. It influences mortality, influences a country's development, and influences the development of children over a long period of time. There is no longer disagreement on these facts because data exist, but now convincing evidence needs to show that nutrition interventions can "deliver" at a reasonable price.

Collaboration

One workshop participant felt that real incentives to work collaboratively are still absent in the international nutrition community. From an institutional perspective, resources for staff and activities are made available by funders if requested by an individual agency. In terms of funding, there seem to be more incentives for acting individually and autonomously than in a coherent way that is fully aligned with what national governments might see as a priority. Another participant disagreed, stating that one of the reasons resources in food and nutrition have been lacking is *because* of the fragmentation. Evidence is beginning to show that alignment is proving attractive to funders.

Coordination Mechanism for Funding

Funding for food and nutrition varies from level to level and subject to subject. One area where the food and nutrition community needs to move toward harmony is on the coordination of what happens with money that goes through multilateral channels. One participant suggested that if governments want to access resources, they currently have to "knock on lots of different doors," present multiple applications, and deal with many different people. A solution (that was credited to Jeff Sachs) is that of a vertical fund that would lump all the donor money into one place, much like the Global Fund to Fight AIDS, Tuberculosis and Malaria.

It was also suggested that a coordinating mechanism is needed to make sure that when money comes from different multilateral and bilateral agencies, particularly at the country level, there is a very clear and transparent process for which national authorities can get information about what money they might be able to access and how they can access it. Additionally, this mechanism would help the multilateral agencies themselves in responding collectively to a national request.

Lessons Learned from Tobacco Control

Obviously the tobacco and food sector are dramatically different. The successes seen through the Framework Convention on Tobacco Control are often referred to, however, in discussions of food and nutrition policy. The tobacco control community recognized the need for UN coherence, and the priorities of the different UN agencies were clarified. For example, when WHO took charge of a UN-wide agreement on the roles of The World Bank, FAO, UNICEF, International Monetary Fund, and WHO, progress began. This had broad implications at the country level, because people saw the importance of defining various agencies' approaches, such as FAO dealing with tobacco supply, while WHO dealt with demand reduction.

A second lesson learned from tobacco control is the importance of getting a stronger voice from local NGOs throughout the process of negotiation. Tobacco control witnessed incredible organization of NGOs who held everyone to account and gave support. In the nutrition world, there are few major indigenous voices from developing countries that are active in the major debates. Part of the problem the nutrition community faces has to do with not hearing the outrage from local voices.

Finally, it should be noted that tobacco control was a legally driven process. There was a regular convening of governments, NGOs, and the private sector (including the pharmaceutical industry) at every major meeting. The fact that trade, health, agriculture, and all of the complex interests were represented meant that common constituencies moved ahead at the same time.

In terms of food and nutrition, part of the problem is in delineating the responsibility of WHO and FAO. Going back to the 1946 constitution, the role of WHO on food and nutrition relative to FAO is very "fuzzy." Additionally, the WFP didn't exist at that time, but it now plays a very strong and critical role, so it also needs to be built into the process.

"Food and Nutrition Security" Tactic

It was suggested that the international nutrition community begin referring to "food and nutrition security" all in one expression. Food security is on the map, but nutrition security and food security are not the same concept, and nutrition policies are somehow seen as separate from food security; if policies and programs for food and nutrition security are discussed jointly, it might make a big impact.

Another participant noted the importance of differentiating between *food safety* and *nutrition* in the context of food security. There is some controversy among FAO membership around the focus on nutrition versus food safety.

REFERENCES

Bhutta, Z. A., T. Ahmed, R. E. Black, S. Cousens, K. Dewey, E. Giugliani, B. A. Haider, B. Kirkwood, S. S. Morris, H. Sachdev, and M. Shekar. 2008. What works? Interventions for maternal and child undernutrition and survival. *Lancet* 371(9610):417-440.

Gates, B. 2008. *Making Capitalism More Creative*. Retrieved January 29, 2009, from http://www.time.com/time/business/article/0,8599,1828069,00.html.

Grebmer, K. V., B. Nestorova, A. Quisumbing, R. Fertziger, H. Fritschel, R. Pandya-Lorch, and Y. Yohannes. 2009. *Global Hunger Index: The Challenge of Hunger: Focus on Financial Crisis and Gender Inequality*. Cologne, Germany: International Food Policy Research Institute.

Levine, R. 2007. *Case Studies in Global Health: Millions Saved*. Sudbury, MA: Jones and Bartlett Publishers.

Levine, R., and D. Kuczynski. 2009. *Global Nutrition Institutions: Is There an Appetite for Change?* Washington, DC: Center for Global Development.

Morris, S. S., B. Cogill, and R. Uauy. 2008. Effective international action against undernutrition: Why has it proven so difficult and what can be done to accelerate progress? *Lancet* 371(9612):608-621.

PepsiCo. 2007. *PepsiCo 2007 Annual Report: Performance with Purpose*. Retrieved February 2, 2010, from http://www.pepsico.com.

Sheeran, J. 2008. Innovating against hunger and undernutrition. *Global Forum Update on Research for Health* 5:174-176.

WHO. 2009. *Nutrition for Health and Development*. Retrieved August 22, 2009, from http://www.who.int/nutrition/en/index.html.
Yach, D. 2008. The role of business in addressing the long-term implications of the current food crisis. *Globalization and Health* 4.
Yunus, M. 2008. *A Poverty-Free World: When? How?* Retrieved January 12, 2009, from http://www.muhammadyunus.org.

7

U.S. Policy in Food and Nutrition

The U.S. government can play an important role in the fight to end global hunger, and there is a renewed sense of political will to address these issues. This chapter covers what is being done to reorient U.S. policy in food and nutrition from the perspectives of the *Roadmap to End Global Hunger*, the U.S. Agency for International Development (USAID), U.S. Department of State, U.S. Department of Agriculture (USDA), and the Chicago Initiative on Global Development. As described by moderator Jackie Judd of the Kaiser Family Foundation, the following presenters discussed what the U.S. government can and should do to help avoid future food crises and mitigate the negative nutritional effects of those that cannot be avoided.

THE ROADMAP TO END GLOBAL HUNGER

James McGovern, B.A., M.P.A.,
Representative for Massachusetts' 3rd Congressional District
U.S. House of Representatives

Although these are interesting and challenging times, the issue of ending hunger must take on a renewed sense of importance and urgency. The United Nations estimates that the number of hungry people in the world is over 1 billion (FAO, 2009). A statistic of this magnitude is difficult to comprehend. The number is so huge that some may lose the human ability to feel it—or some may be overwhelmed and choose to ignore the crisis. The fact is this: there are some issues that cannot be solved in this lifetime, but ending hunger is not one of them. Ending hunger can be achieved with political will. This presentation addresses the *Roadmap to End Global Hunger.* It will discuss how an idea was born, how it turned into the report titled *The Roadmap to End Global Hunger,* and how the recommendations of that report have been translated into legislation, spearheaded by Jo Ann Emerson and James McGovern.

Background

In May 2008, the U.S. Government Accountability Office (GAO) issued a report that described why donor nations, including the United States, were failing in their efforts to help sub-Saharan African nations meet the first Millennium Development Goal of cutting hunger in half by 2015. The authors of this report, Tom Melito and Phil Thomas, briefed the cochairs of the House Hunger Caucus about the report and its findings. One of the central issues that was raised was the lack of coordination and the lack of any clear strategy about how the United States would make an effective contribution to reducing the incidence of hunger and malnutrition in sub-Saharan Africa, or work with those nations on how to create longer-term food security.

This led the House Hunger Caucus to begin discussions about the need for a specially appointed coordinator or office—or a "Hunger Czar"—to oversee a comprehensive, government-wide strategy to address global hunger and food security. This person would be responsible for helping to coordinate the often very uncoordinated food security programs on the ground. The global food crisis of 2008 put into sharp relief how many programs the United States has on food aid, nutrition, and food security and how they are spread over a variety of federal departments, agencies, and jurisdictions. The same problem exists on Capitol Hill, with global food security programs under the jurisdiction of the Agriculture, Foreign Affairs, Ways and Means, and Financial Services, to name just the principal committees.

This led Jo Ann Emerson and James McGovern to lead a crusade for a comprehensive government-wide strategy and for a coordinator on global hunger and food security. The day after he was elected President in 2008, a bipartisan letter was sent to Barack Obama from 116 members of Congress, calling for a comprehensive government-wide strategy and the appointment of a White House coordinator of such a strategy. In addition, meetings were held with Secretary of State Hillary Clinton, as well as members of the Obama transition teams for the Department of State, USAID, and USDA, to discuss the importance of a comprehensive, government-wide strategy that would maximize efforts to reduce global hunger and promote nutrition and long-term food security.

In spring 2008, a diverse group of nongovernmental organizations (NGOs) began talking about drafting a blueprint for the next administration on how U.S. programs and policies could more effectively and successfully address global hunger, nutrition, and food security. The NGOs had their own "jurisdictional" problems, with some focusing mainly on emergency and humanitarian relief operations, while others were engaged in agricultural development, women and children, health and hygiene interventions, research and development, or market development; the list of their various issue, field, and regional expertise goes on and on. After months of discussion, this broad-based coalition found consensus. As a result, in February 2009 the findings and recommendations were released and presented in a report—*The Roadmap to End Global Hunger.*

The Roadmap to End Global Hunger

The *Roadmap* is noteworthy for being simple, straightforward, and brief. It recommends that U.S. government actions to alleviate global hunger and promote food security be the following:

- Comprehensive—It must involve a government-wide effort and integrate all programs.
- Balanced and flexible—U.S. actions must carefully balance and meet emergency needs, longer-term investments in agriculture, and safety nets for the most vulnerable, especially during this global food and financial crisis.
- Sustainable—U.S. actions need to increase the capacity of people and governments to ultimately feed and care for themselves, reduce the impact of hunger-related shocks (whether they are natural or man-made), and be environmentally sustainable and responsive to the new challenges of climate change.
- Accountable—The comprehensive strategy and individual programs need clear targets, benchmarks, and indicators of success. Monitoring and evaluation systems to measure and improve programs need to be developed and implemented.
- Multilateral—Not only should the United States contribute its fair share to the multilateral efforts to address global hunger, nutrition, and food security, but its strategy should also strengthen the multilateral effort and provide international leadership.

The *Roadmap* recommends four basic actions to alleviate hunger and promote food security.

Create a White House Office on Global Hunger

To lead the efforts of this office, the President would appoint a global hunger coordinator. Concretely, the purpose of the office and the coordinator is to create a permanent entity to pull all related federal agencies together and design and carry out a comprehensive government-wide strategy. Equally important, this position holds the backing of the President and ensures accountability that assignments are carried out, determines what is and is not working, what can be improved, and what needs to be eliminated—without regard to turf, budget, or other individual agency interests.

Resurrect the Congressional Select Committee on Hunger

This would allow one central committee—and the *Roadmap* proposes it be bicameral—to focus on issues of hunger, nutrition, and food security.

Required Components of a Comprehensive Strategy to Alleviate World Hunger

The components of a comprehensive federal strategy to alleviate world hunger and promote food security require, first, a well-managed emergency response capability. Second, a comprehensive strategy must include safety nets, social protection, and the reduction of the risk of disaster. Two types of nutrition programs are required: one that focuses on mothers and children, emphasizing comprehensive nutrition before the age of 2; the other incorporates nutrition across the board in all food security programs. Finally, market-based agriculture and the infrastructure necessary for its development must be in place. In all of these areas, the *Roadmap* proposes a special emphasis on and sensitivity to the centrality of women in securing sustainable food security, increased agricultural development and productivity, and the reduction of malnutrition, undernutrition, and hunger.

Specific Recommendations and Funding Targets

The *Roadmap* provides specific recommendations and funding targets across a number of accounts in order to measure whether the administration and its agencies are on track to meet these critical global requirements.

The Roadmap Becomes Legislation

House Resolution 2817, the Roadmap to End Global Hunger and Promote Food Security Act of 2009, was introduced on June 11, 2009. The legislation references the total increased investment of $50.36 billion called for over 5 years, fiscal year (FY) 2010 through FY 2014, for agricultural development, nutrition (including maternal and child programs and for other vulnerable populations), school feeding programs, productive safety net programs, emergency response, research and development, and technical assistance programs. In addition, under this bill, the first increase in agricultural development funding for FY 2010 was targeted at $750 million—and President Obama exceeded that and asked for $1 billion.

The Way Forward

President Obama has designated Secretary of State Hillary Clinton to coordinate a government-wide approach to create, design, and implement a comprehensive U.S. strategy on global hunger, nutrition, agriculture development, and food security. This has been implemented and has been reflected in the President's FY 2010 budget, and the President's announcements at both the G20 meeting in London and the G8 Summit in Italy. The *Roadmap* recommendations and the NGOs that make up the Roadmap Coalition have played a critical role in supporting the U.S. coordinated effort.

A number of organizations and voices all pushing in the same direction for similar priorities, such as the Partnership to End Hunger and Poverty in Africa, are important contributors to the *Roadmap*. The majority of resources will be invested in the areas of greatest need, including Africa and South Asia; however, a number of regions where nations are on the verge of breakthroughs should be included as well. For example, Guatemala and Brazil are carrying out Zero Hunger campaigns, including a special emphasis on ending child hunger. With their leadership, there is a hemisphere-wide initiative to end hunger in the Americas. The United States should be included in this effort and should find ways to support it and contribute to its success.

Emphasis on Nutrition

In June 2009 at the World Food Prize meeting, Secretary Clinton highlighted the seven principles for a food security strategy. The strategy was worrisome because it did not include nutrition. This was different from the comprehensive message presented by Secretary Clinton in a briefing received in April 2009. In addition, nutrition once again failed to have a central role in the announcement at the G8 on agricultural development and global food security.

The emphasis must be made to focus on the under-two population, and the staffing, resources, funds, and coordination must be mobilized for this priority. At the same time, nutrition should be more fully incorporated and emphasized in all antihunger and food security programs. All children need nutritious food, so programs for vulnerable children should fully integrate nutrition into their policies, programs, and projects. Nutritious meals need to be provided to school-age children, and nutrition education must be provided to the children, teachers, parents, and communities served by those schools. Nutrition education should be promoted for pregnant women, in addition to all families and communities that are beneficiaries or touched by such programs. Emergency operations should emphasize nutrition, especially for children of all ages, and use foods that meet the special nutritional and developmental needs of children.

As part of a comprehensive vision, nutrition and food security programs need to integrate the necessary global health interventions into their projects, including deworming, immunizations, vitamin A supplements, and micronutrient fortification, as well as clean water, hygiene, waste management, and even watershed management. That is a comprehensive approach. That is a government-wide approach. That is the *Roadmap*.

USAID'S RESPONSE TO THE FOOD CRISIS AND PREVENTING MALNUTRITION FOR THE FUTURE

Michael Zeilinger, M.D., M.P.H.,
Chief of Nutrition Division, Bureau for Global Health
U.S. Agency for International Development

USAID defines food security as existing when all people at all times have both physical and economic access to sufficient food to meet their dietary needs for a productive and healthy life. There are three main components to food security by USAID's definition—availability, access, and use—and they are each interrelated with the others. Any of these three by itself does not achieve food security. The determinants of food security vary from country to country, region to region, and community to community, but USAID and the U.S. government must strive to achieve food security by addressing the problem comprehensively. Food availability is defined as including sufficient quantities of food from household production, other domestic output, commercial imports, or food assistance. Food access includes adequate resources to obtain appropriate foods for a nutritious diet. This depends on income available to the household, the distribution of income within the household, and the price of food. Food use, or consumption, includes the proper biological use of food, requiring a diet providing sufficient energy and essential nutrients, potable water, and adequate sanitation, as well as knowledge within the household of food storage and processing techniques, basic principles of nutrition, and proper child care and illness management. This presentation will focus on food use or consumption.

Humanitarian Assistance

The global food price crisis began in 2007, and prices peaked in the middle of 2008. In response to the increase and the needs resulting from this crisis, the U.S. Congress provided $1.825 billion to USAID under the President's food security response initiative. This was in addition to existing funds allocated for humanitarian assistance. With these additional funds, USAID focused mostly on addressing the emergency needs of countries that were most affected by the price increase and provided them with humanitarian assistance. The majority of these supplemental funds were used to provide increased emergency food assistance to such countries as Afghanistan, Democratic Republic of Congo, and Ethiopia. USAID also funded local and regional purchases to complement Title

II resources, and it addressed the immediate impact of rising commodity prices on U.S. emergency food aid programs.

In addition, with the geographic focus in Africa, USAID began to implement longer-term programs that addressed the underlying causes of food security. These programs created diversified household assets, improved economic opportunities for the most vulnerable, preserved livelihood access, increased agricultural productivity, promoted seed quality, supported improved management of acute malnutrition and water and sanitation programs, and reduced the risk of disaster through planning and management and improved irrigation techniques.

Development Assistance

In addition to the humanitarian response, $200 million in development assistance was made available by Congress to focus on increasing agricultural productivity in two key regions, East and West Africa. In West Africa, this money was provided to increase agricultural productivity of staple foods, stimulate the supply response, and expand trade of staple foods. In East Africa, the money strengthened the staple food markets to support local and regional procurement. Finally, funding was also provided to the Consultative Group on International Agricultural Research (CGIAR) to disseminate off-the-shelf technologies in sub-Saharan Africa and South Asia.

Food Insecurity

In 2008, there was a crisis when food prices increased drastically, but for most of the countries in which USAID works, food insecurity was already a major problem. It remains a major problem with a higher level of poverty and malnutrition than existed just 2 years ago. The United States has primarily responded to global hunger through humanitarian assistance. While this is essential for reducing the suffering of those devastated by humanitarian disasters, the underlying causes of chronic food insecurity need to be addressed, and food insecurity needs to be eradicated through comprehensive programs that address all three pillars of food security—availability, access, and use (consumption).

Over the past two decades, there has been insufficient progress in reducing hunger. In fact, as the Global Hunger Index map shows, some countries in central and southern Africa have actually increased their index (Figure 7-1).

The Global Hunger Index is based on three indicators: under-five mortality rate, the prevalence of underweight in children, and the proportion of undernourished. Figure 7-2 demonstrates the contribution of the prevalence of underweight to the overall index, particularly in Africa and in South Asia, which are home to 114 million underweight children representing 77 percent of the global burden of underweight (Black et al., 2003). Thus, reducing the prevalence of underweight children under 5 years is critical to reducing overall hunger. Focusing on nutri-

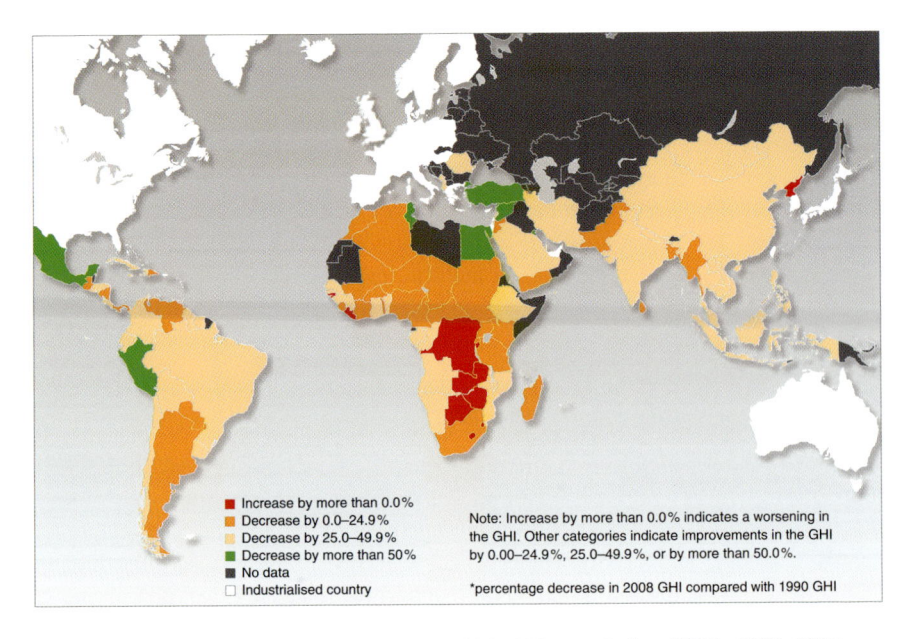

FIGURE 7-1 Progress toward reducing the Global Hunger Index (GHI), 1990–2008.
NOTE: Increase by more than 0.0% indicates a worsening in the GHI. Other categories indicate improvements in the GHI by 0.00–24.9%, or by more than 50%. Percentage decrease in 2008 GHI compared with 1990 GHI.
SOURCE: von Grebmer et al., 2008. Reproduced with permission from the International Food Policy Research Institute, http://www.ifpri.org. The original report can be found online at: http://www.ifpri.org/sites/default/files/pubs/cp/ghi08.pdf.

tion as part of a comprehensive food security strategy can help lift families out of poverty, improve economic growth, increase educational attainment, and can even reduce maternal and child mortality (Figure 7-2).

In Bangladesh, about 8 million children are undernourished despite relative availability of food (UNICEF, 2008). There, malnutrition is the cause of two out of three child deaths. It is also the cause of long-term damage to cognitive development, physical growth, and productivity for millions more. The damage caused by malnutrition to physical growth, brain development, pregnancy, and early childhood is irreversible. It leads to permanently reduced cognitive function and diminished physical capacity throughout adulthood.

A family's nutrition status is perpetuated from generation to generation (Figure 7-3). Malnourished girls are more likely to have children who are malnourished, who will be less productive workers, earn lower wages, and who will develop noncommunicable diseases later in life. In countries like Bangladesh where half the women and children suffer from malnutrition, the losses are staggering. Yet, this entire process is preventable.

U.S. POLICY IN FOOD AND NUTRITION 143

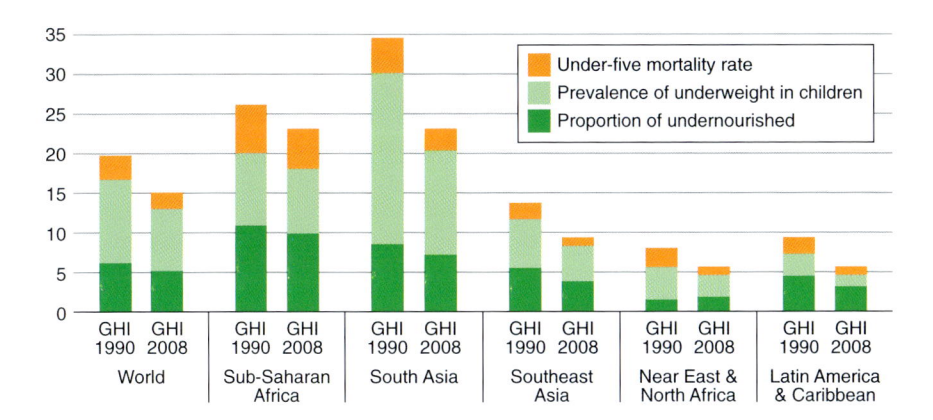

FIGURE 7-2 Contribution of three indicators (under-five mortality rate, prevalence of underweight in children, and proportion of undernourished) to the Global Hunger Index (GHI), 1990 and 2008.
SOURCE: von Grebmer et al., 2008. Reproduced with permission from the International Food Policy Research Institute, http://www.ifpri.org. The original report can be found online at: http://www.ifpri.org/sites/default/files/pubs/cp/ghi08.pdf.

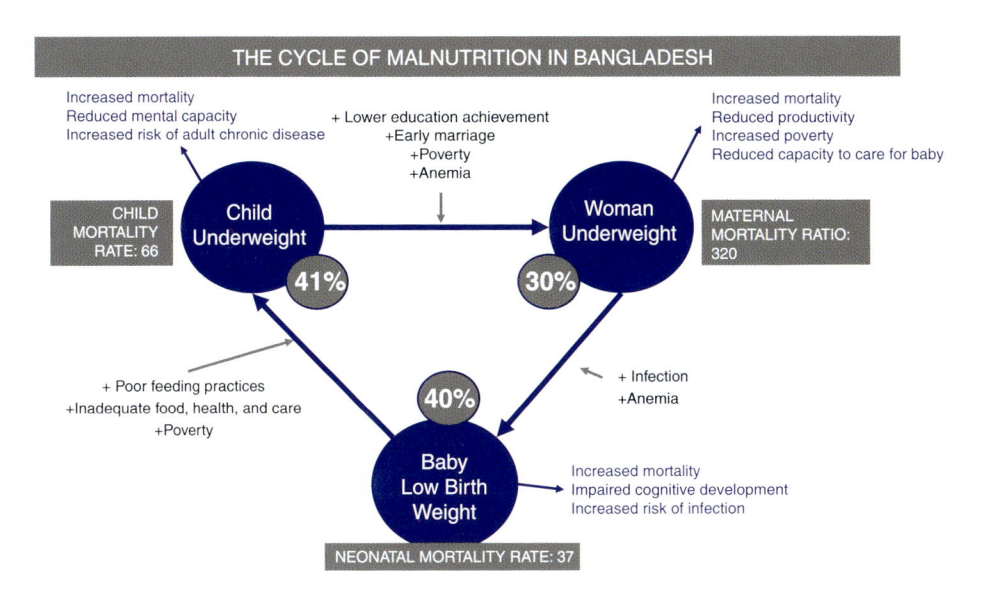

FIGURE 7-3 The cycle of malnutrition in Bangladesh.
SOURCE: USAID, 2009.

USAID's Approach to Food Security

USAID is currently engaged in the process of strengthening its approach to food security and is transitioning to a balanced approach that offers both humanitarian actions for emergency purposes as well as developmental assistance that will increase agricultural productivity, increase trade, support women, and increase support for families, among other components. The support for women and families includes a strong focus on reducing and preventing malnutrition.

The principles that are guiding USAID's work in nutrition in the Global Health Bureau, include:

- A focus on the chronically hungry,
- The window of opportunity is from pregnancy to 24 months,
- The quality of foods and their use within the household are crucial elements of food security,
- Prevention of malnutrition is ultimately the most sustainable approach, and
- Programs should be country owned and designed based on the country-specific determinants of malnutrition and food insecurity.

Malnutrition can be prevented. Evidence-based interventions exist to improve nutrition through integrated community-based approaches, effective targeting, and public-private partnerships. These interventions include the following:

- Community-based education and counseling programs promote maternal nutrition, exclusive breast-feeding under 6 months, and the introduction of appropriate locally available complementary foods for children aged 6 to 23 months.
- Micronutrients are provided for the most vulnerable, including vitamin A for children under 5 years, iron for women and children, and iodized salt.
- Community management of acute malnutrition is integrated into national health services and community outreach.
- Fortification programs are part of value-chain development with the private sector, including biofortification.
- Innovative foods for young children are delivered in partnership with the private sector, including ready-to-use complementary foods.
- Access to and consumption of diverse and high-quality foods is increased by promoting community, school, and kitchen gardens.

FOOD SECURITY IN THE 21ST CENTURY

Nina Fedoroff, Ph.D., Science and Technology Adviser to the Secretary of State and to the Administrator of the U.S. Agency for International Development

Global crises abound, and each of them rapidly pushes aside the previous. The current financial crisis has added to last year's global food and energy crises and even the ever-looming climate change crisis. This presentation will address the U.S. Department of State's approach to food security and will begin by addressing the changes in the earth's climate and how these changes have profound implications for food security.

Climate Change and Food Security

The earth has warmed over the past 100 years by almost 1°C. By now it is no longer contentious to assert that this is anthropogenic.[1] The atmosphere's carbon dioxide concentration has increased by one-third since 1750, predominantly due to the burning of fossil fuel, but also because of deforestation. More importantly, the rate of increase is somewhere between 100 and 1,000 times faster than the rates observed historically.

How will climate change affect the crops that feed the world? The average temperature in Europe in the summer of 2003 was 3.6°C higher than average, although rainfall was normal. That summer, between 30,000 and 50,000 people died from heat-related causes, a statistic that received much attention. But the effect on crops received little notice: Italy saw a 36 percent decrease in maize yield, while France experienced a 30 percent reduction in maize and fodder yields, a 25 percent decrease in fruit harvests, and a 21 percent reduction in wheat yields. Summer temperatures this high will become more frequent in the coming decades, and by mid-century, such record high temperatures are likely to be the norm. Familiar crops do not survive well at these temperatures, and even a brief period of very high temperatures at the critical time of flowering and pollination can devastate a crop. Temperature also influences how fast a plant develops and reaches maturity. Higher temperatures speed plants through their developmental phases. Annual crops like corn, wheat, and rice set seed just once and then stop producing. As temperature increases from the crop's optimum, the growing period is shorter. Both the shortening of the growing period and the decrease in photosynthetic efficiency at higher temperatures reduce yields.

Irrigation can be used to cool crops at critical times, however many countries are already unsustainably overpumping aquifers. These include three of the big-

---

[1] Caused or produced by humans.

gest grain producers—China, India, and the United States. More than half the world's people live in countries where water tables are falling.

The primary options to address these issues include more efficient agriculture, particularly with respect to water use; intensive crop breeding; modern molecular genetic modification for both drought and heat tolerance, as well as insect and disease resistance; and development of new crops. Some of this can be done immediately; some will require research and changes in public opinion.

Overall, investments in agricultural development have declined. Historically, rural poverty decreased and agricultural productivity increased with (1) better education, (2) new technologies, and (3) investment. It is essential that the recent trend toward providing food aid at the expense of investing in agricultural research and development be reversed.

Unfortunately, many well-meaning people around the world today believe that genetically modified crops are dangerous. What this means for the present is that modern science cannot be used to improve crops in many countries, including most of Africa. Although the U.S. regulatory apparatus is not completely prohibitive, it is dauntingly complex and so expensive that public-sector researchers have largely turned away from molecular crop improvement. Progress needs to be made in moving toward a regulatory framework that is based on actual risks and real scientific evidence, not hypothetical risks and popular fears.

U.S. Department of State Effort on Food Security

Secretary Clinton sees food security as one of the areas in which the State Department can make a difference. An interagency committee has been working to lay out a vision, goals, and strategy. The vision is straightforward—the United States envisions a world in which all people have reliable access to safe, nutritious, and affordable food. The goals are familiar and very much in line with the Millennium Development Goals. The goals are to:

- Halve the proportion of young children who are undernourished;
- Halve the proportion of people who suffer from hunger;
- Halve the proportion of women and men living on less than $1.25 a day; and
- Build sustainable agriculture systems that create jobs, increase incomes, and raise agricultural productivity without harming the environment.

The State Department's strategy has seven pillars:

1. Increase farm and farmer productivity.
2. Stimulate postharvest private-sector growth.
3. Enable private-sector investment and development.
4. Increase trade flows.

5. Support women and families in agriculture.
6. Use natural resources sustainably.
7. Support research and development of agricultural technologies and expand access to knowledge and training.

The substance of the first pillar is expanding access to land, seeds, fertilizer, and irrigation, with an emphasis on women's access, as well as expanding extension services, training and credit, and working with social entrepreneurs. The second is about transportation networks, storage facilities, and food processing, as well as local procurement, transport, and distribution of emergency food aid. The third is about creating private-sector markets, streamlining business regulations, addressing land tenure, and supporting the development of local organizations to increase participation in decision making. The fourth pillar of the strategy is about developing regional markets, lowering trade barriers, helping to develop food safety standards, improving market information and communications systems, and improving access to finance for agricultural trade and agribusiness development. The fifth is adapting services and training to the needs of women, increasing their access to credit, financial services, education, and land ownership, as well as improving child nutrition through school feeding programs. The sixth is about promoting sustainable agricultural practices and adapting agriculture to climate change. The seventh is about increasing support for the CGIAR; facilitating public-private partnerships, as well as collaborative partnerships between U.S. universities and universities in sub-Saharan Africa, Asia, and Latin America; and creating a competitive award fund to support research.

The overall strategy relies on the development of country-led plans that bring all stakeholders to the table, as well as the coordination of multilateral support through the UN High-Level Task Force on Food Security. At present, the State Department policy group is engaging with other agencies to develop a "whole-of-government" approach.

USDA'S RESPONSE TO THE CRISES AND FUTURE PERSPECTIVES

Rajiv Shah, Under Secretary and Chief Scientist
U.S. Department of Agriculture

This presentation discusses lessons learned from the U.S. experience and what that means as the country goes forward to try to integrate agriculture with nutrition. Although this is not the first time the global community has come together to try to significantly reduce hunger and poverty in an agriculture-focused manner, it is perhaps the best chance to succeed.

Background

An estimated 1.1 billion people live on $1 a day, and a large portion of that population is dependent on agriculture for its basic economic opportunities and its basic access to foods (The World Bank, 2009). Agriculture is a central component of global extreme poverty. It is not the entire solution to all poverty, but it is simply the largest slice of the most extreme poverty.

A number of people in the world suffer from hunger, which is defined as not having enough basic calories to meet energy needs in a very basic and subsistent manner. A few years ago, this was estimated at 840 million (FAO, 2006), and in 2008 it increased to 960 million (FAO, 2008). It is now reported that there are more than 1 billion hungry people in the world (FAO, 2009). This is a number that is very difficult to comprehend. In addition, 10 million children die each year because they have poor nutrition that makes them prone to diarrheal and infectious diseases that they do not have the physical capacity to survive (Black et al., 2003). These are complex and interrelated problems.

The U.S. Experience

The U.S. experience offers a template for how to potentially address the problem of hunger elsewhere. A long time ago, the United States was a largely agrarian society that struggled to provide enough food to meet the needs of its population. The country was hyperdependent on and responsive to the rain-fed agriculture that was variable, challenging, and difficult. The land-grant university system was developed by Abraham Lincoln in the 1860s, and it started to systematically invest in agricultural research. Coupled with that system, an extension service was created, which today is the most dramatic way to deliver information to farmers that has ever been created. It is the largest adult education program in the country. Basic extension services are the primary means for developing public-sector electronic content that reaches farm communities and rural communities through the e-extension service. Today, 4-H programs are able to reach more children than any other structured program in the federal government—about 6 million children participate in 4-H programs each year.

The coupling of research and extension helped increase U.S. productivity. In addition to these, a third critical component is the developing of structured markets. In the United States, markets have existed and improved over time because of the phytosanitary[2] standards that allow commodities to be traded and that allow people to understand what buyers are buying and what sellers are selling. Systems have been in place to ensure the protection of the food supply and our food safety. Such basic market mechanisms often do not exist in the countries that face the biggest burden of hunger.

---

[2] Of agricultural goods crossing borders; sanitary with regard to pests and pathogens.

Nutrition Assistance

The largest percentage of the USDA budget is committed to nutrition assistance programs. Such programs target children, vulnerable populations, and communities that would not otherwise have access to basic food items and would have to spend a very high percentage of their disposable income acquiring food. The United States has a low-cost food supply in which only 11 percent of the average American's disposable income is spent on food. In the countries with the most vulnerable populations, approximately 70–80 percent of total disposable income is spent on acquiring food. In that environment, when food prices increase, it presents difficult challenges to some very vulnerable populations. The nutrition assistance component of having a robust, active, and self-sustaining agriculture system is an important component that is often overlooked.

Experience of China and India

Lessons can be learned from other countries, including China and India. Over the past few decades, both countries have pursued effective agricultural development strategies and have significantly increased their agricultural productivity through the Green Revolution and crop varieties, along with a range of other policy interventions. China has been far more successful at translating those agricultural productivity gains into reductions in the rates of malnutrition; India has been far less successful. In India, there was less investment in lagging regions—those parts of the country that did not experience the increases in productivity and income—and were essentially left behind as a result. In addition, India lacked infrastructure in nutrition support for rural communities. Even today, a disjointed system exists where there are significant and persistent rates of rural malnutrition and rural poverty, despite having a significant growth rate that created a modern middle class in India. These experiences can provide lessons as new food security strategies are implemented in a range of countries. It will be important to couple nutrition-oriented efforts with agriculture programs in order to maximize their impact.

New Technologies

Research and science can offer unique contributions to the intersection of agriculture and nutrition. Several technologies exist that can improve basic human nutrition in a targeted way for children and women. The first is biofortification. This approach takes the staple foods that some of the poorest populations depend on and enhances them with the micronutrients that women and children need, like vitamin A, zinc, and folic acid.

One example is the orange-fleshed sweet potato. A joint project between the International Food Policy Research Institute (IFPRI) and HarvestPlus bred higher amounts of beta carotene into sweet potato varieties in parts of Africa. This project caused a lot of debate and concerns about whether it would actually make a difference in children's health. Data now demonstrate that introducing a sweeter, orange-fleshed product into the daily food supply has doubled the serum retinol[3] in vitamin A–deficient children in northern Mozambique and northern Uganda. The initiative was successful because the product tastes good and children like it. A great deal of investment in agricultural technology development or agricultural foods and productivity has not taken into account the preferences of the people. For biofortification initiatives to be successful, it is important to allow farmers to taste different varieties that are created, and then target efforts on those traits the people want most. This type of research is important and must target the customers. For example, if the customers are malnourished children under 5 years, see what they like, and use that feedback to inform the research.

Another example of a scientific contribution is vegetable breeding. A number of programs are focused on breeding improved varieties of vegetables, while also making sure they are pest and disease resistant. Such research systems exist in the United States and are a reason that the United States is the world leader in agriculture. People in lower-income countries suffering from food insecurity need the same types of research systems to protect their own ability to produce food, particularly vegetables, because they help achieve dietary diversity and improve micronutrient sufficiency in families.

The third area of technological promise concerns livestock. As families' incomes grow from $2 a day to $10 a day, some of the first luxury items that are brought into the market basket of goods they consume are meat products and dairy products—and thus higher levels of protein. Yet there has been very little effective investment in dairy productivity, livestock improvement, genetic improvement for animals, or pasteurization technologies that would allow smallholder farmers (who don't have access to large chilling plants or other types of energy systems) to protect their product from spoiling in order to sell it into a more formal market. These technologies are being developed and need to reach small farmers through food security initiatives.

Conclusion

Without extension services, education programs, and the targeting of women, technology alone will not solve the problem of hunger and undernutrition. Technology, however, does hold tremendous untapped potential. The USDA is working to expand its resources in technology development, phytosanitary standards, and market standards, as well as its ability to regulate and develop commercial systems for commodity exchanges. All of those solutions can help to achieve both the agriculture objectives and the health and nutrition objectives that are so

---

[3] Biochemical indicator of vitamin A deficiency.

critical in helping millions of low-income children and families lead healthy and productive lives in some of the toughest places on earth.

RENEWING AMERICAN LEADERSHIP IN THE FIGHT AGAINST GLOBAL HUNGER AND POVERTY

Catherine Bertini, Professor of Public Administration
Syracuse University
Dan Glickman, Chairman and CEO
Motion Picture Association of America, Inc.

This presentation describes the objectives and recommendations of the Chicago Council on Global Affairs' Chicago Initiative on Global Agricultural Development as laid out in its report, *Renewing American Leadership in the Fight Against Global Hunger and Poverty*, 2009.

Background

In June 2008, work began on the Chicago Initiative with a grant from the Bill & Melinda Gates Foundation. The objective of the Chicago Initiative was to generate political, media, and public support for U.S. international leadership to put agricultural development back at the center of U.S. development policy. Food aid remains a top priority, and while the United States should play a critically important role with food aid, U.S. programs should build on food aid and supplement it with additional funding. The United States should have a food security program and a program to support agriculture that can reach the problems where they begin.

The Chicago Council on Global Affairs held a number of meetings with opinion leaders, campaigns, and members of Congress. By the end of 2008, the group had developed a white paper to present to the new administration and met with various transition teams to give them the outline of what was proposed in the plan.

In his inaugural speech, President Obama pledged to work alongside the people of poor nations to help make their farms flourish. In February 2009, the Chicago Council's report was made public, Catherine Bertini and Daniel Glickman testified before the Senate Foreign Relations Committee, and President Obama pledged $1 billion to agricultural development. There were a number of similarities between the Chicago Initiative, the Global Food Security Act, and President Obama's plan—especially the emphasis on agriculture education and research, infrastructure, policy coordination, and strengthening USAID in public-private partnerships (Table 7-1).

TABLE 7-1 Comparison of Three Versions of U.S. Food Aid

	Chicago Council on Global Affairs	Global Food Security Act	President Obama's Plan
Goal	Mobilize U.S. knowledge, training, assistance, and investment to increase the productivity and income of smallholder farmers.	Alleviate poverty and enhance human and institutional capacity by investing in agricultural research and education.	Improve the lives of poor populations by focusing on agricultural growth in rural communities.
Agricultural education and research	Increase USAID-sponsored scholarships, U.S. land-grant universities' partnerships, and agricultural research at CGIAR.	Collaborate with U.S. land-grant universities to increase support for agricultural research including genetically modified technology.	Work with U.S. land-grant universities to expand development and use of modern technology and strengthen host country research institutions.
Rural and agricultural infrastructures and market access	Increase support for rural and agricultural infrastructure, especially in sub-Saharan Africa, working with international financial institutions.	Improve agricultural infrastructure, finance and markets, safety net programs, job creation, and household incomes.	Strengthen national and regional trade and transport corridors to boost farmers' access to seeds, fertilizers, rural credit, and markets.
Policy coordination	Create Council on Global Agriculture and a deputy in National Security Council.	Create a Special Coordinator for Global Food Security within the Executive Office.	N/A
USAID's role	Strengthen the leadership and in-house capacity of USAID.	Designate USAID as the lead agency.	USAID to develop a comprehensive Food Security Initiative.
Encourage public-private-partnerships	Yes	Yes	Yes
Cost	$341.05 million (first year) $1.03 billion (fifth year)	$750 million (FY 2010) $2.5 billion (FY 2014)	$1 billion (FY 2010)

Chicago Initiative on Global Agricultural Development Report Recommendations

A strategic plan for combating global hunger and poverty must involve agricultural development, emergency food assistance, nutrition, agricultural research, extension, and investment in modern methods of agriculture. The Chicago Initiative provided the following recommendations in its report:

- Increase support for agricultural education.
- Increase support for agricultural research.
- Increase support for rural infrastructure.
- Improve the national and international institutions that deliver agricultural assistance.
- Revise U.S. agricultural policies.

Agricultural Education

The Chicago Initiative recommends increased USAID support for sub-Saharan African and South Asian students studying agriculture; increased American agricultural university partnerships with universities in sub-Saharan Africa and South Asia; direct support for agricultural education, research, and extension for young women and men; a special Peace Corps cadre of agriculture training and extension volunteers; and support for primary education for rural girls and boys through school feeding programs.

Agricultural Research

This recommendation aims to increase support for agricultural research in sub-Saharan Africa and South Asia. This includes providing external support for agricultural scientists working in national agricultural research systems, research conducted at the international centers of CGIAR, collaborative research between scientists from sub-Saharan Africa and South Asia at U.S. universities, and creating a competitive award fund to provide an incentive for high-impact agricultural innovations to help poor farmers.

Rural Infrastructure

The Chicago Initiative recommends increased support for rural and agricultural infrastructure, especially in sub-Saharan Africa. This recommendation also encourages a revival of The World Bank lending for agricultural infrastructure in sub-Saharan Africa and South Asia. In addition, it recommends accelerating disbursal of the Millennium Challenge Corporation funds already obligated for rural infrastructure projects.

Institutional Reform

This recommendation is to improve the national and international institutions that deliver agricultural development assistance. It recommends restoring the leadership role of USAID; rebuilding USAID's in-house capacity to deliver and administer agricultural development assistance programs; improving interagency coordination for America's agricultural development assistance efforts; strengthening the capacity of the U.S. Congress to collaborate in managing agricultural development assistance policy; and improving the performance of international agricultural development and food institutions, notably FAO.

Policy Reform

This Chicago Initiative recommendation aims to improve U.S. policies currently seen as harmful to agricultural development abroad. There is a need to improve America's food aid policies, repeal current restrictions on agricultural development that might lead to more agricultural production for export, review USAID's long-standing objection to any use of targeted subsidies to reduce the cost to poor farmers, revive international negotiations aimed at reducing trade-distorting policies (including agricultural subsidies), and adopt biofuel policies that place a greater emphasis on market forces and on the use of nonfood feed stocks.

Next Steps

The Chicago Initiative highlights a number of issues and makes recommendations on them; however, it is important to note that the main challenge is prioritizing these issues and providing leadership within the U.S. government to nonprofits, universities, and international NGOs. The next steps include:

- Expansion of G20 initiatives,
- Passage of the Global Food Security Act,
- Strengthening leadership of USAID,
- Continued agricultural development advocacy activities,
- A strategy for leveraging donors and international organizations,
- Continued in-depth discussion of key agricultural development issues, and
- Creativity of Obama appointees.

This is a collective responsibility; now that the U.S. government has made these issues a priority, and the G8 has placed them on the agenda, it will take additional pressure to develop the strategic plans and the leadership to make sure the goals are reached.

DISCUSSION

This discussion section encompasses the question-and-answer sessions that followed the presentations summarized in this chapter. Workshop participants' questions and comments have been consolidated under general headings.

Coordinating U.S. Global Health Initiatives

A number of U.S. initiatives that currently exist could naturally be blended, such as food and nutrition programs with HIV/AIDS programs. It was noted that the administration is working to integrate the various agencies and programs into one comprehensive plan. A wide variety of issues are part of a worldwide global health initiative; however, food and agricultural issues do not attract the same interest as a number of other issues. For example, AIDS and malaria initiatives get much support worldwide, especially when compared to the international hunger effort, whose political resonance is not nearly as developed or as widespread. It is positive to note that governments, NGOs, and foundations are beginning to take hold of this issue and develop the political will to make these issues an important part of global development efforts.

Political Will

Sustaining political will is an important element to defeating global hunger and undernutrition. How can this be sustained in an environment in which so many important issues are on the table? Policy makers are faced with a multitude of simultaneous issues, so the key is to find leaders who are willing to sustain the effort and articulate the problems in a way people can understand. For example, in the area of food assistance, the public *does* recognize the moral and humanitarian aspect, which has helped gain support for food aid. However, agricultural assistance has fallen off the map and needs to be restored. People have become disengaged with trying to "help people help themselves" in the world of agriculture. It was noted that because food security is now on the President's agenda, support is growing because people realize that food and economic issues affect everyone—not just the NGO and university communities, but the American private sector and corporate world as well.

The Role of Food Aid

The view was expressed that the United States should not lead its fight against global hunger with food aid (although food aid should not be decreased and more flexibility should be allowed for local purchase of food). Food aid becomes more important in emergency situations when there is no other food available for purchase, or where buying food in the market would greatly disrupt the market and increase prices in others areas. In the United States, there are some important constituencies involved in food aid, including commodity groups and shippers, that make reform of food aid challenging.

U.S. Leadership

The role of the United States as a leader in coordinating the number of organizations that work in the nutrition landscape was discussed. It was noted that a coordination role is more important than a commanding role. It is important for the United States to listen to what works in other countries and help guide sustainable food security solutions according to the local context.

It was noted that in the field, although there are competing interests, there is actually much more synergy among various agencies and groups than is seen in Washington, DC. Typically, the countries with the worst hunger and development indicators actually have *more* synergy within the U.S. overseas missions. In these situations, those on the ground see what needs to be done and are able to put aside their differences to improve the dire situation in that country.

WORKSHOP CLOSING REMARKS

Reynaldo Martorell, Ph.D., Robert W. Woodruff Professor of Public Health
Rollins School of Public Health, Emory University

In reflecting on the workshop, Dr. Martorell drew attention to the fact that the title of the workshop, *Mitigating the Nutritional Impacts of the Global Food Price Crisis*, spoke about that price crisis alone. As the planning committee began preparations for the workshop, however, the financial crisis occurred and the planning committee decided to expand the workshop's focus to include both the food price and the economic crisis in the workshop proceedings.

During the 3-day workshop, many topics were covered by various presenters. Per Pinstrup-Andersen gave a detailed analysis of the global food price crisis; he highlighted the fact that it was not only the high prices for commodities, but also the price volatility that caused a great deal of problems. Dr. Pinstrup-Anderson further pointed out that prior to the food crisis, commodity prices were too low to provide a good livelihood for producers and to encourage investment in agriculture. In this way, low prices are also a problem.

Hans Timmer from the World Bank spoke about the financial crisis. Not only were the countries that participate actively in financial markets impacted by this crisis, but also the countries of sub-Saharan Africa, for example, were also very much impacted. While the price of commodities did decrease during the financial crisis, prices on average remained higher than before the food price crisis. It is important to note that governments' capacities to respond with safety nets are weakened because of the financial crisis. It is therefore crucial for the global economy—and in the best interests of the United States—that developing countries begin to have healthy economies; this is part of the solution for the global economy.

Marie Ruel discussed the fact that vulnerable people in urban areas were expected to be affected most severely because they are net buyers of food. In reality, though, a substantial number of people in rural areas also seem to be gravely affected. The country experiences described by the case studies demonstrated the fact that the data are not ideal, and a variety of nutrition surveillance mechanisms were discussed. The session that examined the global response made clear that the crisis did not affect every country in the same way and there was tremendous diversity in response. In many cases, governments' reactions exacerbated the problem by imposing export bans and interfering with trade. Among the negative nutritional effects that were described in the presentations were deteriorations in dietary quality when people eat less of more expensive foods; typically more expensive foods are the more nutrient dense foods such as meat, vegetables, and fruits.

The food price and financial crises present an opportunity to rethink approaches to food security and nutrition, to coordinate and deploy systems in a better way, to motivate policy makers at national and global levels to a greater commitment to action, and to do effective advocacy. There was much discussion about forging partnerships with the private sector, NGOs, and civil society, and to advocate for an increased level of resources. Workshop participants talked about a way forward, which includes increasing investment in agriculture, increasing productivity, and bringing science and technology to bear.

The Way Forward—Themes from the Workshop

The following themes emerged during the workshop through several speakers' presentations and during discussion sessions with workshop participants. These themes are not intended to be and should not be perceived as a consensus of the participants, nor the views of the planning committee, the IOM, or its sponsors.

- The current crisis presents an opportunity to motivate donors and engage affected country governments in efforts to address undernutrition, hunger, and food insecurity in vulnerable populations.
- There is a window of opportunity with women and children where known nutritional interventions will be most effective and have a long-term payoff, as described in the 2008 *Lancet* series on maternal and child undernutrition.
- There is a simultaneous call for better quality data to inform program design and effectiveness, but there is also a critical need to immediately move forward with proven programs and policies to mitigate hunger and undernutrition in vulnerable populations.

- Short-term, emergency actions are not sufficient to remedy recurring food crises; instead, both short- and long-term investments in global food and agriculture systems are needed.
- Mechanisms to help vulnerable populations cope with food price volatility and to prevent future shocks are required.
- It is important to draw upon the expertise of governments, NGOs and civil society, the private sector, foundations, and the broad spectrum of actors in the international nutrition and agriculture sectors.
- The roles of the multiple UN agencies that work to promote the food and nutrition security of vulnerable populations need to be clarified.
- Fostering engagement with the private sector may yield new expertise and resources.
- A stronger voice from indigenous NGOs is needed. Such local NGOs could benefit from capacity-building efforts to encourage ownership and political involvement.

REFERENCES

Black, R., S. Morris, and J. Bryce. 2003. Where and why are 10 million children dying every year? *Lancet* 361(9376):2226-2234.

Food and Agriculture Organization of the United Nations. 2006. *The State of Food Insecurity in the World.* Rome: Food and Agriculture Organization.

———. 2008. *Number of Hungry People Rises to 963 Million.* Rome: Food and Agriculture Organization.

———. 2009. *1.02 Billion People Hungry.* Rome: Food and Agriculture Organization.

Renewing American Leadership in the Fight Against Global Hunger and Poverty: The Chicago Initiative on Global Agricultural Development. 2009. Chicago: The Chicago Council on Global Affairs.

The World Bank. 2009. *Food Crisis: What The World Bank is Doing.* Retrieved November 9, 2009, from http://www.worldbank.org/foodcrisis/bankinitiatives.htm.

UNICEF. 2008. *The State of the World's Children 2009: Maternal and Newborn Health.* New York: United Nations Children's Fund.

USAID. 2009 (unpublished). *The Cycle of Malnutrition in Bangladesh.*

von Grebmer, K., H. Fritschel, B. Nestorova, T. Olofinbiyi, R. Pandya-Lorch, and Y. Yohannes. 2008. *The Challenge of Hunger: The 2008 Global Hunger Index.* Bonn: Welthungerhilfe; Washington, DC: International Food Policy Research Institute; Dublin: Concern Worldwide.

Appendix A

Workshop Agenda

MITIGATING THE NUTRITIONAL IMPACTS OF THE GLOBAL FOOD PRICE CRISIS
JULY 14–16, 2009, WORKSHOP AGENDA

The Kaiser Family Foundation
Barbara Jordan Conference Center
1330 G Street, N.W., Washington, DC 20005

Day One: Tuesday, July 14, 2009

8:30	Continental breakfast available
9:00–9:10	Welcome **Harvey Fineberg**, Institute of Medicine President
9:10–9:20	Introduction **Reynaldo Martorell**, Emory University, Workshop Moderator
9:20–9:30	Welcome from sponsor **Ellen Piwoz**, Global Health Program, Bill & Melinda Gates Foundation

SESSION 1
The Dual Crises: Tandem Threats to Nutrition

Session Objectives: To set the stage for the deliberations by having an overview of the recent food price crisis and how it, in tandem with the current economic crisis (to be discussed July 15 by Hans Timmer), affects developing countries.

Moderator: Reynaldo Martorell, Workshop Planning Committee Chair

9:30–10:00	The Recent and Current Food Price Crisis and Future Perspectives **Per Pinstrup-Andersen**, Cornell University
10:00–10:30	Question and answer session
10:30–10:45	Break

SESSION 2
Impacts on Nutrition

Session Objectives:	To understand the pathways from the food price and economic crises to nutritional impact, including a discussion of existing evidence and vulnerable populations.

Moderator: Isatou Jallow, World Food Programme

10:45–11:15	Conceptual Presentation on Pathways to Nutritional Impact **Ricardo Uauy**, London School of Hygiene and Tropical Medicine; University of Chile
11:15–11:45	Existing Evidence of Nutritional Impacts **Francesco Branca**, World Health Organization
11:45–12:15	Are the Urban Poor Particularly Vulnerable? **Marie Ruel**, International Food Policy Research Institute
12:15–1:00	Open discussion
1:00–1:45	Lunch provided

SESSION 3
Responding to the Crises at the Country Level

Session Objectives:	To understand the range of country experiences with the food price and economic crises and their impact on food security and nutrition, as well as country-level responses to these crises.

Moderator: Ruth Oniang'o, Rural Outreach Program, Kenya

1:45–2:05	The Role of Ministries in Responding to the Crises at the Country Level **Ruth Oniang'o**, Rural Outreach Program, Kenya
2:05–2:35	Review of National Responses to the Food Crisis **Hafez Ghanem**, Food and Agriculture Organization

Country Experiences and Responses: Case Studies

2:35–2:55	The Case of Mexico **Graciela Teruel Belismelis**, Iberoamericana University
2:55–3:15	The Global Food Price Crisis and Food Development Strategy in China **Fangquan Mei**, State Council Food and Nutrition Consultant Committee; Chinese Association for Agricultural Modernization
3:15–3:30	Break
3:30–3:50	Food Prices, Consumption, and Nutrition in Ethiopia: Implications of Recent Price Shocks **Paul Dorosh**, International Food Policy Research Institute, Ethiopia
3:50–4:10	Bangladesh Case Study **Josephine Iziku Ippe**, United Nations Children's Fund (UNICEF)
4:10–5:15	Moderated discussion
5:15	Adjourn for the day

Day Two: Wednesday, July 15, 2009

8:30	Continental breakfast available
9:00–9:05	Introduction to day two **Reynaldo Martorell**, Emory University

SESSION 4
Revisiting the Dual Crises: Tandem Threats to Nutrition

Session Objectives: To set the stage for the deliberations by having an overview of the current economic crisis and how it, in tandem with the recent food price crisis (addressed July 14 by Per Pinstrup-Andersen), affects developing countries.

Moderator: Reynaldo Martorell, Emory University

9:05–9:35 The Current Economic Crisis and Future Perspectives
Hans Timmer, The World Bank

9:35–10:05 Question and answer session

SESSION 5
A Role for Nutrition Surveillance in Addressing the Global Food Crisis

Session Objectives: To encourage a broad discussion of nutrition surveillance, including existing nutrition surveillance systems, their capacity to monitor food price fluctuations, and the gaps and needs for improved surveillance.

Moderator: Keith West, Johns Hopkins Bloomberg School of Public Health

10:05–10:35 Nutrition Surveillance in Relation to the Food Price and Economic Crises
John Mason, Tulane University

10:35–10:50 Break

10:50–11:50 Strengths and Limitations of Past, Existing, and Budding Nutrition Surveillance Systems (15 minutes/speaker)

Andrew Thorne-Lyman, Helen Keller International Nutrition Surveillance Projects; Harvard School of Public Health
Chris Hillbruner, FEWS NET; Chemonics
Anna Taylor, Listening Posts Project; Save the Children UK
Ellen Mathys, FANTA-2 Project; Academy for Educational Development

11:50–12:50	Moderated discussion
12:50–1:30	Lunch provided

SESSION 6
The Global Response to the Crises

Session Objectives:	To understand the landscape of the global nutrition field, those who work in it, and their respective roles and capacities to respond to the food price and economic crises.
	Moderator: Hans Herren, Millennium Institute
1:30–2:00	Introduction to the Global Nutrition Landscape **Ruth Levine**, Center for Global Development
2:00–3:00	The Role and Capacity of Civil Society, the Private Sector, and Foundations in Responding to the Crises (15 minutes/speaker) **Haddis Tadesse**, Gates Foundation **Derek Yach**, PepsiCo **Asma Lateef**, Bread for the World **Tom Arnold**, Concern Worldwide
3:00–3:15	Break
3:15–4:30	The Role and Capacity of UN Agencies in Responding to the Crises (15 minutes/speaker) **David Nabarro**, United Nations Task Force on Global Food Security Crisis **Werner Schultink**, United Nations Children's Fund (UNICEF) **Martin Bloem**, World Food Programme **Hafez Ghanem**, Food and Agriculture Organization **Francesco Branca**, World Health Organization
4:30–5:30	Moderated discussion
5:30	Adjourn for the day

Day Three: Thursday, July 16, 2009

8:30	Continental breakfast available
9:00–9:05	Introduction to day three **Reynaldo Martorell**, Emory University

SESSION 7
Reorientation of U.S. Policy in Food and Nutrition

Session Objectives: To discuss what the U.S. government can and should do to help avoid future food crises and to mitigate the negative nutrition effects of those that cannot be avoided.

Moderator: Jackie Judd, Kaiser Family Foundation

9:05–10:45	**The U.S. Government Response to the Crises** (20 minutes/speaker) **Representative James McGovern**, Roadmap to End Global Hunger **Michael Zeilinger**, U.S. Agency for International Development **Nina Fedoroff**, U.S. Department of State **Rajiv Shah**, U.S. Department of Agriculture **Catherine Bertini and Dan Glickman**, Chicago Initiative on Global Agricultural Development
10:45–12:00	Moderated discussion
12:00–12:15	Summary discussion and wrap-up **Reynaldo Martorell**, Emory University
12:15	Adjourn

Appendix B

Speaker Biographies

Tom Arnold, M.S., is Chief Executive of Concern Worldwide. Concern Worldwide is Ireland's largest nongovernmental organization (NGO) helping with emergencies, long-term development, and advocacy, and it is active in 30 countries, mainly in Africa and Asia. Prior to working with Concern Worldwide, Mr. Arnold was Assistant Secretary General and Chief Economist in the Irish Department of Agriculture and Food. He was Chairman of the Organisation for Economic Co-operation and Development (OECD) Committee of Agriculture from 1993 to 1998 and Chairman of the Working Group on Agricultural Policies and Markets from 1990 to 1993. At an earlier stage in his career, he worked with the European Commission for 10 years, 3 of which were in Africa. Mr. Arnold has been appointed to a number of international bodies in recent years. He was a member of the advisory board of the United Nations (UN)'s Central Emergency Response Fund (CERF) (2006–2009), a member of the UN Millennium Project's Hunger Task Force (2003–2005), and a member of the World Economic Forum Expert Group on poverty and hunger. He was a member of the Irish government's Hunger Task Force (2007–2008). He is currently Chairman of the European Food Security Group, a network of 40 European NGOs working to enhance food security in developing countries. Mr. Arnold is a member of the Advisory Board for the International Food Policy Research Institute's (IFPRI's) 20/20 Initiative, which seeks to develop and promote a shared vision and consensus for action for assuring sustainable food and nutrition security for all by 2020. He is currently a member of the Irish government's Commission on Taxation and a governor of the *Irish Times*, Ireland's leading newspaper. Mr. Arnold is a graduate in agricultural economics from University College Dublin and has master's degrees from the Catholic University of Louvain and Trinity College Dublin.

Catherine Bertini, B.A., served as Executive Director of the UN World Food Programme from 1992 to 2002, turning the organization into the world's largest humanitarian aid agency, and was the UN Under-Secretary General for Management from 2003 to 2005. Ms. Bertini's responsibilities as the Under-Secretary General included controlling the United Nation's $3 billion biennial budget as well as human resources and security for 9,000 staff members. For her innovative leadership in assisting hundreds of millions of victims of war and natural disaster throughout the world, Ms. Bertini received the 2003 World Food Prize. She has also been honored by the Republic of Italy with its Order of Merit and by the Association of African Journalists with its Prize of Excellence. In 1996, *The Times of London* named her one of "The World's Most Powerful Women."

Martin Bloem, M.D., Ph.D., is Chief for Nutrition and HIV/AIDS Policy, UN World Food Programme, in Rome, Italy. He holds a medical degree from the University of Utrecht and a doctorate from the University of Maastricht, and he has joint faculty appointments at both Johns Hopkins Bloomberg School of Public Health and Tufts University Friedman School of Nutrition Science and Policy. Dr. Bloem has had more than two decades of experience in nutrition research and policy. He was a Medical Officer at the Ministry of Defense in The Hague, the Netherlands, and at CIVO-TNO Toxicology & Nutrition Institute, TNO in Zeist, the Netherlands. He was a scientific consultant at the Nutrition Supplement Cooperation Project in Thailand. He has served as Country Director of Helen Keller International (HKI) Bangladesh and HKI Indonesia; as Regional Director for HKI Asia Pacific; and as Senior Vice President and Chief Medical Officer of HKI Singapore.

Francesco Branca, M.D., Ph.D., is Director of the Department of Nutrition for Health and Development at the World Health Organization (WHO). He has been Regional Advisor on Nutrition and Food Security at the WHO Regional Office for Europe, dealing with the design and implementation of nutrition policy and nutrition surveillance. He has been a senior scientist at the Italian Food and Nutrition Research Institute in Italy, dealing with nutrition surveillance; with the design, management and evaluation of public health nutrition programs in Africa and Central Asia; and the implementation of research on the biological effects of micronutrients and bioactive compounds in food. He has been a member of the European Food Safety Authority Panel on Nutrition, the President of the Federation of the European Nutrition Societies, and has taught public health nutrition at the University of Rome. Dr. Branca has been working extensively in primary health care in the Horn of Africa.

Paul Dorosh, B.A., M.A., Ph.D., is Senior Research Fellow with the International Food Policy Research Institute (IFPRI) and Program Leader of the Ethiopia Strategy Support Program in Addis Ababa, Ethiopia. Before moving to Ethiopia, he worked 6 years as a Senior Economist with The World Bank, in the Spatial and Local Development Team and in the South Asia Agricultural and Rural Development Unit. He earlier worked for 6 years as a Senior Research Fellow with the International Food Policy Research Institute, including 4 years in Dhaka, Bangladesh, where he was an Advisor to the Ministry of Food. From 1989 to 1997, Dorosh was a Senior Research Associate and Associate Professor with Cornell University where he worked on the effects of structural adjustment policies and poverty in Madagascar, Niger, and other countries of sub-Saharan Africa. He holds a B.A. in applied mathematics from Harvard University and a Ph.D. in applied economics from the Food Research Institute of Stanford University.

Nina V. Fedoroff, Ph.D., is Science and Technology Adviser to the Secretary of State and to the Administrator of the U.S. Agency for International Development (USAID). Dr. Fedoroff is on leave from the Pennsylvania State University, where she is the Willaman Professor of Life Sciences and the Evan Pugh Professor in the Biology Department and the Huck Institutes of the Life Sciences. Dr. Fedoroff is a leading geneticist and molecular biologist who has contributed to the development of modern techniques used to study and modify plants. She received her Ph.D. in molecular biology from the Rockefeller University in 1972. In 1978, she became a staff member at the Carnegie Institution of Washington and a faculty member in the Biology Department at Johns Hopkins University. In 1995, Dr. Fedoroff joined the faculty of the Pennsylvania State University, where she served as the founding director of the Huck Institutes of the Life Sciences. Dr. Fedoroff has done fundamental research in the molecular biology of plant genes and transposons, as well on the mechanisms plants use to adapt to stressful environments. Her book, *Mendel in the Kitchen: A Scientist's View of Genetically Modified Foods*, published in 2004 by the Joseph Henry Press of the National Academy of Sciences, examines the scientific and societal issues surrounding the introduction of genetically modified crops. Dr. Fedoroff is a member of the National Academy of Sciences, the American Academy of Arts and Sciences, and the European Academy of Sciences. She has served on the National Science Board of the National Science Foundation, and she is a 2006 National Medal of Science laureate.

Harvey V. Fineberg, M.D., Ph.D., is the President of the Institute of Medicine. He served as Provost of Harvard University from 1997 to 2001, following 13 years as Dean of the Harvard School of Public Health. He has devoted most of his academic career to the fields of health policy and medical decision making. His past research has focused on the process of policy development and implementation, assessment of medical technology, evaluation and use of vaccines, and dissemination of medical innovations. Dr. Fineberg helped found and served as president of the Society for Medical Decision Making and also served as consultant to the World Health Organization. At the Institute of Medicine, he has chaired and served on a number of panels dealing with health policy issues, ranging from AIDS to new medical technology. He also served as a member of the Public Health Council of Massachusetts (1976–1979), as chairman of the Health Care Technology Study Section of the National Center for Health Services Research (1982–1985), and as president of the Association of Schools of Public Health (1995–1996). Dr. Fineberg is coauthor of the books *Clinical Decision Analysis*, *Innovators in Physician Education*, and *The Epidemic That Never Was*, an analysis of the controversial federal immunization program against swine flu in 1976. He has coedited several books on such diverse topics as AIDS prevention, vaccine safety, and understanding risk in society. He has also authored numerous articles published in professional journals. Dr. Fineberg is the recipient of several honorary degrees and the Joseph W. Mountin Prize from the U.S. Centers for Disease Control and Prevention (CDC). He earned his bachelor's and doctoral degrees from Harvard University.

Hafez Ghanem, M.A., Ph.D., was appointed Assistant Director-General, Economic and Social Development Department in November 2007. Mr. Ghanem holds a B.A. and an M.A. in economics (economic development and international trade) from the American University in Cairo, Egypt, and a Ph.D. in economics (trade, econometrics) from the University of California, Davis. Mr. Ghanem began his professional career with the World Bank in Washington, DC, in 1983. Between 1995 and 1997, he served as Principal Economist for Armenia, Georgia, Moldova, and Ukraine. From 1997 to 2000, he worked successively as Sector Leader, Public Economics and Trade Policy for Europe and Central Asia Region. Subsequently, he was appointed as the World Bank's Country Director in Madagascar, Comoros, Mauritius, and Seychelles. Since 2004, he served as Country Director in Nigeria.

Daniel Glickman, B.A., J.D., is chairman and CEO of the Motion Picture Association of America, Inc., (MPAA), which serves as the voice and advocate of the American motion picture, home video, and television industries. Its members include Walt Disney Studios Motion Pictures, Paramount Pictures, Sony Pictures Entertainment, Twentieth Century Fox Film Corp., and NBC Universal and Warner Bros. Entertainment. Prior to joining the MPAA, Mr. Glickman was the director of the Institute of Politics at Harvard University's John F. Kennedy School of Government (2002–2004). He also served as Senior Advisor to the law firm of Akin Gump Strauss Hauer & Feld in Washington, DC. Mr. Glickman served as U.S. Secretary of Agriculture from March 1995 until January 2001. Under his leadership, the department administered farm and conservation programs; modernized food-safety regulations; forged international trade agreements to expand U.S. markets; and improved its commitment to fairness and equality in civil rights. Before his appointment as Secretary of Agriculture, Mr. Glickman served for 18 years in the U.S. House of Representatives, representing the 4th

Congressional District of Kansas. During that time, he was a member of the House Agriculture Committee, including 6 years as chairman of the subcommittee with jurisdiction over federal farm policy issues. Moreover, he was an active member of the House Judiciary Committee, Chairman of the House Permanent Select Committee on Intelligence, and a leading congressional expert on general aviation policy. Before his election to Congress in 1976, Glickman served as president of the Wichita, Kansas, school board; was a partner in the law firm of Sargent, Klenda and Glickman; and worked as a trial attorney at the U.S. Securities and Exchange Commission. He received his B.A. in history from the University of Michigan and his J.D. from The George Washington University. He is a member of the Kansas and the District of Columbia bars. Mr. Glickman serves on the board of directors of the Chicago Mercantile Exchange, Hain-Celestial Corp., Communities in Schools, Food Research and Action Center (FRAC), the National 4-H Council, the William Davidson Institute, and the Center for U.S. Global Engagement. He is also a member of the Genocide Prevention Task Force, chaired by former Secretaries Madeleine Albright and Bill Cohen; the Council on Foreign Relations; and the Kansas Bioscience Authority. In addition, Mr. Glickman serves as the Chicago Council on Global Affairs Cochair of the Global Agriculture Development Project (with Catherine Bertini). He is a former member of the international advisory board of the Coca-Cola Co. He has been a Senior Fellow and part-time instructor in the public policy departments at Georgetown University and Wichita State University and is a lecturer on public policy at Harvard University's John F. Kennedy School of Government.

Hans R. Herren, Ph.D. *(Planning Committee Member)*, has been the President of Millennium Institute since May 2005. Prior to joining Millennium Institute, he was Director-General of the International Center for Insect Physiology and Ecology (ICIPE) in Nairobi, Kenya. He also served as Director of the Africa Biological Control Center of the International Institute of Tropical Agriculture (IITA) in Benin. At ICIPE, Dr. Herren developed and implemented programs in the area of human, animal, plant, and environmental health (the 4-H paradigm) as they relate to insect issues. At IITA, he conceived and implemented the highly successful biological control program that saved the African cassava crop, averting Africa's worst-ever food crisis. He serves on the boards of numerous organizations, including as Cochair of the International Assessment of Agricultural Knowledge, Science, and Technology; Chairman of BioVision, a Swiss foundation with a global mandate to alleviate poverty and improve the livelihoods of poor people while maintaining the precious natural resource base that sustains life; President of the International Association of the Plant Protection Sciences; and former member of the National Research Council Board on Agriculture and Natural Resources. Dr. Herren earned his Ph.D. at the Federal Institute of Technology in Zurich, Switzerland, and holds numerous awards that recognize his distinguished and continuing achievements in original research.

Christopher Hillbruner, M.S., received a master's degree in food policy from the Tufts University Friedman School of Nutrition Science and Policy in 2007. He now works at Chemonics as a Food Security Warning Specialist for USAID's Famine Early Warning Systems Network (FEWS NET). FEWS NET collaborates with international, regional, and national partners to provide timely and rigorous early warning and vulnerability information on emerging and evolving food security issues. FEWS NET professionals in Africa, Central America, Haiti, Afghanistan, and the United States monitor and analyze relevant data and information in terms of its impacts on livelihoods and markets to identify potential threats to food security. Once these issues are identified, FEWS NET uses a suite of communications and decision support products to help decision makers act to mitigate food insecurity.

Josephine Iziku Ippe, M.Sc., has 15 years of experience in nutrition and related programs in Africa, mostly in east and central Africa. She is currently the Nutrition Manager in the UNICEF Bangladesh Country Office, based in Dhaka. Before joining UNICEF, she worked for 5 years with Oxfam GB in Kenya, Uganda, Tanzania, Congo, Nicaragua, and South Sudan. During her fifth year with Oxfam GB, she also provided technical support to countries mainly within the Horn and east and central African regions. She was a Nutrition Programme Manager for SCF-UK in Darfur, Sudan, and later she was a Nutrition Coordinator for Action African in Need (AAIN) in the Southern Sudanese Refugees Camp, Ikafe, and North Uganda. She was an Assistant Project Officer of Nutrition for UNICEF/ Operation Life Line (OLS). She managed the emergency nutrition program in the Malawi Country Office as a Nutrition Project Officer, and then became the Chief of the Nutrition Section for UNICEF, North Sudan, responsible for emergency and development nutrition projects and sector coordination. She has a B.Sc. in family sciences from the Ahfad University for Women in Omdurman, Sudan, and an M.Sc. in human nutrition from the London School of Hygiene and Tropical Medicine, University of London.

Isatou Jallow, M.Sc. *(Planning Committee Member)*, is a nutritionist and gender advocate with 22 years of field and policy experience at both the country and international level. She currently serves as the Chief of Women, Children, and Gender Policy of the UN World Food Programme based in Rome. Some of her achievements include the transformation of the Ministry of Health Nutrition Unit of The Gambia to a National Nutrition Agency under the Office of the Vice President, with a mandate to coordinate nutrition across the various sectors. She also adapted the global UNICEF/WHO Baby Friendly Hospital Initiative to a community initiative (Baby Friendly Community Initiative) incorporating maternal and infant nutrition, environmental sanitation, and personal hygiene into 10 steps for communities to implement. The initiative was initially piloted in 12 communities in The Gambia and gradually scaled up to almost 300 communities. Ms. Jallow is currently Cochair of the UN Standing Committee on Nutrition (SCN) working group on Breastfeeding and Complementary Feeding. She is also an advisory committee member of the IFPRI-coordinated "Millions Fed" project. She was selected as the second Abraham Horwitz lecturer for the UN SCN 28th session in 1998. In May 2009, she was awarded the Medal of National Order of The Republic of The Gambia for her achievements in nutrition. Ms. Jallow holds a M.Sc. in nutrition from the University of Oslo, Norway.

Jackie Judd is Vice President and Senior Advisor for Communications at the Kaiser Family Foundation. She joined the foundation in 2003 as a Senior Visiting Fellow. Ms. Judd's current responsibilities include managing and developing multimedia projects for the foundation's events and websites, involvement in the foundation's international partnerships related to the coverage of HIV/AIDS and managing the Technology Working Group. Ms. Judd is a former long-time broadcast journalist covering a range of issues including politics, health care policy, and Congress. She was with ABC News for 16 years as a correspondent for *World News Tonight with Peter Jennings*, *Nightline*, and *Good Morning America*. At National Public Radio, she was a news anchor and cohost on *Morning Edition* and *All Things Considered*. Ms. Judd is also a former CBS News radio correspondent. Her honors include National Emmy awards, an Edward R. Murrow Award, the Joan Barone Award, the David Bloom award, a duPont Award, a commendation from Women in Radio and Television for a series on women's health issues, and an Overseas Press Club Citation of Excellence. She received a bachelor's degree from American University (AU) in 1974. Ms. Judd serves on the Dean's Advisory Committee at AU's School of Communications, and she is a member of the board of directors of Rebuilding Together of Washington, DC.

Asma Lateef, B.A., M.A., is Director of Bread for the World Institute. An affiliate of Bread for the World, a nonpartisan, Christian antihunger organization, the institute publishes its annual *Hunger Report*, as well as briefing papers and educational materials on issues related to domestic and international hunger and poverty. She has also worked as a consultant at the UN Conference on Trade and Development and the International Labour Organization. She holds a bachelor's degree from McGill University in Montreal, Canada, a postgraduate diploma from the London School of Economics, and a master's degree in economics from the University of Maryland.

Ruth Levine, Ph.D., is Vice President for Programs and Operations and Senior Fellow at the Center for Global Development (CGD), where she leads the center's work on global health policy. Dr. Levine has a doctorate in economic demography from Johns Hopkins University. She is a health economist with more than 15 years of experience designing and assessing the effects of social sector programs in Latin America, eastern Africa, the Middle East, and South Asia. Before joining the CGD, Dr. Levine designed, supervised, and evaluated loans at the World Bank and the Inter-American Development Bank. Between 1997 and 1999, she served as the advisor on the social sectors in the office of the executive vice president of the Inter-American Development Bank. She has co-authored *The Health of Women in Latin America and the Caribbean* (The World Bank, 2001), *Millions Saved: Proven Successes in Global Health* (CGD, 2004, updated as *Cases in Global Health: Millions Saved* [Jones and Bartlett, 2007]), and *Performance Incentives for Global Health: Potential and Pitfalls* (CGD, 2009).

Reynaldo Martorell, Ph.D. *(Planning Committee Chair)*, is the Robert W. Woodruff Professor of Public Health in the Rollins School of Public Health at Emory University, where he also served as Chair of the Department of Global Health. Previously, he was a Professor in the Division of Nutritional Sciences at Cornell University and at the Food Research Institute at Stanford University. Dr. Martorell's research interests include maternal and child nutrition (particularly in developing countries), child growth and development, the significance of early childhood malnutrition for short- and long-term human function, micronutrient malnutrition, and the emergence of obesity and chronic diseases in developing countries. Dr. Martorell's policy interests include global health concerns, particularly programs and policies in food and nutrition, issues dealing with hunger and malnutrition, and the health implications of changes in diet and lifestyles in developing countries (including the emergence of obesity and related chronic diseases of dietary origin in developing countries). He was active at the National Research Council during the 1980s, serving on the Food and Nutrition Board, its Committee on International Nutrition Programs, and the Subcommittee on Vitamin A Deficiency Prevention and Control. More recently he chaired the IOM Planning Committee for the Joint U.S.–Mexico Workshop on Preventing Obesity of Children and Youth of Mexican Origin. Dr. Martorell is a consultant to The World Bank, UNICEF, and WHO; past President of the Society for International Nutrition Research; and past Associate Editor of the *Journal of Nutrition*. Dr. Martorell received a bachelor's degree in anthropology from St. Louis University and a Ph.D. in biological anthropology from the University of Washington. He was elected to the Institute of Medicine in 2002 and currently serves on the Committee on Childhood Obesity Prevention and as a member of the Food and Nutrition Board.

John Mason, Ph.D., is a Professor in the Department of International Health and Development at Tulane University. He works to improve nutrition, particularly of children and women, in developing countries. Starting in nutritional biochemistry, Dr. Mason moved on to research child health and nutrition in east and west Africa, before joining the Food and Agricultural Organization where he worked on nutritional surveillance and program planning. He then became Director of the Cornell Nutritional Surveillance Program, conducting research and training both in Cornell and overseas; at the same time he codirected a joint program with UNICEF to promote nutrition in eastern and southern Africa. Returning to the UN in 1986, Dr. Mason was Technical Secretary of the UN Coordinating Committee on Nutrition (ACC/SCN) based in WHO, where he started the series of UN reports on the world nutrition situation, the Refugee Nutrition Information System, and supervised 15 UN publications on nutrition policy issues. He joined Tulane University in 1996. His interests are currently focused on nutrition policy development; on approaches to sustaining community-based programs for nutrition improvement; and micronutrient deficiencies, in terms of epidemiology and prevention. Currently he is acting as an advisor to The World Bank and UNICEF for the National Nutrition Program in Ethiopia; to the UN-SCN in reporting on the world nutrition situation; and to WHO as a member of the Expert Reference Group for the Health and Nutrition Tracking System, and for a landscaping exercise to identify priorities for nutrition interventions. Dr. Mason received his bachelor's, master's, and doctoral degrees from University of Cambridge.

Ellen Mathys, M.P.H., is the Senior Food Security Early Warning and Response Specialist with the USAID-funded Food and Nutrition Technical Assistance Project II (FANTA-2). She has 13 years of experience working in 19 low-income countries in the areas of nutrition and food security. From 2004 to 2006 she served as the Livelihoods Advisor with the USAID-funded Famine Early Warning Systems Network (FEWS NET) in Washington, DC. Recently she assisted FANTA and Office of Food for Peace (FFP) with developing guidance on early warning and response, including trigger indicators and guidance on food assistance programming in urban emergencies for the private voluntary organizations (PVO) community. She has developed global tools and best practice guidance for the UN and the Global Nutrition Cluster related to international nutrition and food security assessment in emergencies as well as integrating food/nutrition and HIV programs in crisis and refugee settings. She has an M.P.H. from Tulane University, with a focus on international nutrition and food security and complex emergencies and disaster management.

James McGovern, B.A., M.P.A., is currently serving his seventh term in Congress, and he was first sworn in as U.S. Representative for Massachusetts' 3rd Congressional District in January 1997. Representative McGovern is Vice Chairman of the House Rules Committee, which sets the terms for debate and amendments on most legislation, and he is a member of the House Budget Committee. Representative McGovern is also Cochair of both the Tom Lantos Human Rights Commission and the House Hunger Caucus. Before his election to Congress, Representative McGovern spent 14 years working as a senior aide for the late U.S. Representative John Joseph Moakley (D-South Boston), former dean of the Massachusetts delegation and Chairman of the House Rules Committee. During those years, Representative McGovern earned a strong reputation as a champion of human rights. In 1989, Representative McGovern was chosen by Congressman Moakley to lead a congressional investigation into the murders of six Jesuit priests and two lay women in El Salvador. In Congress, Representative McGovern has championed several education initiatives, including a bill to increase grant assistance for college students and their families. He has led the fight to provide adequate health care, including home health care, and he has worked to increase funding for the Land and Water Conservation Fund. He has fought to preserve and strengthen Social Security, and he has secured millions of dollars in federal assistance to central and southeastern Massachusetts. Representative McGovern earned his bachelor of arts (1981) and master's of public administration (1984) degrees from the American University, working his way through college by serving as an aide in the office of U.S. Senator George McGovern (D-SD). He went on to manage Senator McGovern's 1984 presidential campaign in Massachusetts, and he delivered his nomination speech during the 1984 Democratic National Convention in San Francisco.

Fangquan Mei, M.S., Ph.D., is the Standing Vice President of the State Food and Nutrition Consultant Committee of the State Council of China. He is also President of the Chinese Association for Agricultural Modernization and the Honorary Director-General of the Agricultural Information Institute, CAAS. He is engaged in research for food and agriculture development strategies, consumption and production structures, and information management. At present, Professor Mei is in charge of key projects on food security and early warning systems in China and the development of an agricultural research information system (ARIS) in China. He is the former President of the Asian Federation for Information Technology in Agriculture. He won the first and second National Science and Technology Progress Award three times, has published 117 papers, is in charge of writing and compiling 13 copies of monographic works, and trained 76 postgraduates (Ph.D. and M.S.). Recently, he was invited to visit 20 countries and present at 68 international academic conferences, including the International Conference on Nutrition (ICN), the FAO 50th Anniversary, the World Fertilizer Conference, the World Food Production Conference, and the Asia Conference on Agricultural Information Technology.

David Nabarro, M.D., serves as an Assistant Secretary General in the United Nations. He holds the position of Senior UN System Coordinator for Avian and Human Influenza, reporting to the UN Deputy Secretary-General, on secondment from the World Health Organization (WHO) since September 2005. In April 2008 he was given an additional responsibility as Deputy UN System Coordinator by the UN Secretary-General, and he has been Coordinator for the Global Food Security Crisis as of January 2009. A physician and public health specialist, Dr. Nabarro has worked in the UK National Health Service, taught at the London and Liverpool Schools of Tropical Medicine, worked in child health programs in

Nepal, served as regional manager for the UK Save the Children Fund in South Asia, and in 1989 served as health and population adviser to the British Overseas Development Administration (ODA) in Nairobi, Kenya. In 1990 Dr. Nabarro moved to London as ODA's Chief Health and Population Adviser and was promoted to the position of Director for Human Development in the Department for International development in 1997. In 1999, he joined WHO to manage the Roll Back Malaria project; in 2000 he became Executive Director in the Office of the Director-General. He transferred to WHO's Sustainable Development and Healthy Environments cluster in 2003. He was then appointed Representative of the Director General for Health Action in Crises, coordinating support for health assessments and aspects of crises preparedness, response, and recovery operations in a variety of locations including Darfur, Liberia, and countries affected by the 2004 Indian Ocean earthquake and tsunami.

Ruth K. Oniang'o, Ph.D. *(Planning Committee Member)*, is the Executive Director of the Rural Outreach Program and has been elected a member of Parliament of the Government of Kenya. Dr. Oniang'o has done much consultation work, such as with the Food and Agriculture Organization (FAO) in nutrition. Her areas of research and consultation are household food and nutritional security, women's nutrition, child health, and community-level agro-processing, in which she has published widely. Dr. Oniang'o has been awarded the Silver Star Medal by the President of the Republic of Kenya for outstanding service to the country in community development through action research in 1995 and the Distinguished Service Medal for national service in 1998. She has served on the board of the Kenya Bureau of Standards, Egerton University Council, and Poverty Eradication Commission. She is the founder and editor-in-chief of the *African Journal of Food, Agriculture, Nutrition, and Development*. She is on the board of the Kenya Gatsby Charitable Trust, Food Security and Sustainable Development Division of the Economic Commission for Africa, Institute for Policy Analysis and Research, International Fertilizer Development Center, Biotechnology Advisory Council of Monsanto-USA, and the Private Sector Corporate Governance Trust. She is also on the advisory committee of the Biofortification Project of the International Food Policy Research Institute. She is a member of the working group forming the Society of African Journal Editors, a panel member of the World Cancer Research Fund International, and Founder-President of the Kenya Union of Food Science and Technology. Dr. Oniang'o received her B.S. (with distinction) and M.S. degrees from Washington State University in the United States and her Ph.D. from the University of Nairobi, Kenya, and she has since spent 20 years in academia.

Per Pinstrup-Andersen, Ph.D. *(Planning Committee Member)*, is the H. E. Babcock Professor of Food and Nutrition Policy at Cornell University and Professor of Development Economics at Royal Veterinary and Agricultural University in

Denmark. Before assuming his current positions, Dr. Pinstrup-Andersen was the Director General of the International Food Policy Research Institute (IFPRI) from 1992 to 2002. He had previously been Director of the Cornell Food and Nutrition Policy Program, Professor of Food Economics at Cornell University, member of the Technical Advisory Committee of the Consultative Group on International Agricultural Research, Research Fellow and Director of the Food Consumption and Nutrition Policy Program at IFPRI, Agricultural Economist at the International Center for Tropical Agriculture (CIAT), Director of the Agro-Economic Division at the International Fertilizer Development Center, and an Associate Professor of the Danish Veterinary and Agricultural University in Copenhagen. He is currently a member of the NRC Roundtable on Science and Technology for Sustainability and has served on other IOM or NRC committees, including the Committee on Agricultural Biotechnology, Health, and the Environment; Committee on International Nutrition Programs; and the Food and Nutrition Board. He holds a B.S. in agricultural economics from the Royal Veterinary and Agricultural University in Denmark and an M.S. and Ph.D. from Oklahoma State University. Dr. Pinstrup-Andersen is the recipient of the 2001 World Food Prize.

Ellen G. Piwoz, Sc.D., M.H.S., is a Senior Program Officer in the Global Health Program of the Bill & Melinda Gates Foundation, responsible for nutrition in the Integrated Health Solutions Development division. Prior to joining the Gates Foundation in 2007, Dr. Piwoz was the Director of the Center for Nutrition at the Academy for Educational Development in Washington, DC. During her 12 years at the academy, she directed the Sustainable Approaches to Nutrition in Africa Project, and she was a senior nutrition advisor to USAID's Africa Bureau Office of Sustainable Development. Dr. Piwoz held adjunct appointments at the Johns Hopkins University Bloomberg School of Public Health and at the University of North Carolina Gillings School of Global Public Health where she was an investigator in several studies examining ways to prevent HIV transmission during breast-feeding. She was an elected councilor of the Society for International Nutrition Research, American Society of Nutrition from 2002 to 2006.

Juan A. Rivera, Ph.D. *(Planning Committee Member),* is the Founding Director of the Center for Research in Nutrition and Health at the National Institute of Public Health and is a Professor of Nutrition in the School of Public Health in Mexico. He is also an Adjunct Professor in the Division of Nutritional Sciences at Cornell University and at the Rollins School of Public Health at Emory University. Dr. Rivera's research interests include the epidemiology of stunting, the short- and long-term effects of undernutrition during early childhood, the effects of zinc and other micronutrient deficiencies on growth and health, the study of malnutrition in Mexico, and the design and evaluation of programs to improve nutritional status of children. Dr. Rivera is a former Director of Nutrition and

APPENDIX B
177

Health at the Nutrition Institute of Central America and Panama. He is Cochair of the International Zinc Nutrition Consultative Group. Since 2000, he has been a member of the panel of experts of the World Cancer Research Fund and has been appointed to the National Academy of Medicine and to the Mexican Academy of Sciences in Mexico. He was a member of the board of the International Union of Nutritional Scientists from 2001 to 2005 and of the Global Alliance for Improved Nutrition Board until 2005. He is Chair of the International Nutrition Council of the American Society for Nutrition. Dr. Rivera has published more than 130 scientific articles, book chapters, and books, and he is currently a member of the Latin American Nutrition Society, the American Society for Nutritional Sciences, and the Society for International Nutrition Research. He served on the IOM Planning Committee for the Joint U.S.–Mexico Workshop on Preventing Obesity of Children and Youth of Mexican Origin. Dr. Rivera earned both his master's and doctorate degrees from Cornell University in international nutrition with minors in epidemiology and statistics.

Marie T. Ruel, Ph.D., is Director of the Poverty, Health, and Nutrition Division at the International Food Policy Research Institute (IFPRI). From 1996 until her current appointment, Dr. Ruel served as Senior Research Fellow and Research Fellow in that division. Since joining IFPRI, she led the Multi-Country Program on Challenges to Urban Food and Nutrition and the Global Regional Project on Diet Quality and Diet Changes of the Poor. Prior to IFPRI, she was head of the Nutrition and Health Division at the Institute of Nutrition of Central America and Panama/Pan American Health Organization (INCAP/PAHO) in Guatemala. Dr. Ruel has worked for more than 20 years on issues related to policies and programs to alleviate poverty and child malnutrition in developing countries. She has published extensively in nutrition and epidemiology journals on topics such as maternal and child nutrition, food-based strategies to improve diet quality and micronutrient nutrition, urban livelihoods, food security and nutrition, and the development of indicators of child feeding and care practices. Dr. Ruel has served on various international expert committees, such as the National Academy of Sciences and the International Zinc in Nutrition Consultative Group. Dr. Ruel received her Ph.D. in international nutrition from Cornell University and her master's in health sciences from Laval University in Canada.

Werner Schultink, M.D., is the Chief of Child Development and Nutrition for the United Nations Children's Fund (UNICEF). UNICEF has worked from its founding on nutrition programming aimed at fulfilling every child's right to adequate nutrition. UNICEF is committed to scaling up and sustaining coverage of its current high-impact nutrition interventions in the program areas of infant and young child feeding, micronutrients, nutrition security in emergencies, and nutrition and HIV/AIDS.

Rajiv J. Shah, M.D., was recently nominated by President Barack Obama as Under Secretary of Research, Education and Economics (REE) and Chief Scientist at the U.S. Department of Agriculture. The REE mission area provides the science that federal agencies, policy makers, researchers, and others draw on to meet challenges facing America's food and agriculture system. The four REE agencies are the Agricultural Research Service (including the National Agricultural Library), Economic Research Service, National Agricultural Statistics Service, and Cooperative State Research, Education, and Extension Service. Dr. Shah was formerly Director of the Agricultural Development Program at the Bill & Melinda Gates Foundation. The foundation's agriculture programs represent a multibillion dollar global effort to reduce hunger and poverty. Dr. Shah joined the foundation in 2001 and served as Director of Strategic Opportunities and Deputy Director of Policy and Finance for Global Health. Before joining the foundation, Dr. Shah was a health care policy advisor on the Gore 2000 presidential campaign and a member of Governor Ed Rendell's (D-PA) transition committee on health. He is a cofounder of Health Systems Analytics and Project IMPACT for South Asian Americans. He also served as a policy aide in the British Parliament and worked at the World Health Organization. Originally from Detroit, Michigan, Dr. Shah earned his M.D. from the University of Pennsylvania Medical School and his M.S. in health economics from the Wharton School, University of Pennsylvania. He is a graduate of the University of Michigan and the London School of Economics. In 2007, Dr. Shah was named a Young Global Leader by the World Economic Forum.

Haddis Tadesse, M.P.A., currently works as Policy and External Relations Officer at the Bill & Melinda Gates Foundation. He is responsible for managing the global development external relation efforts in Africa. Prior to joining the foundation, Mr. Haddis served as the Senior Policy Advisor to Mayor Greg Nickels of Seattle, Washington. His responsibilities included providing policy advice to the mayor, deputy mayor, and the cabinet of Seattle City government on matters of human services, public health, housing, civil rights, and immigrant and refugee affairs. Prior to his position as Senior Policy Advisor, Mr. Tadesse devoted his work to the city of Seattle, becoming a human resource, planning, and development specialist and later the Boards and Commissions Administrator. He also worked as an accounting specialist at the University of Washington Harborview Medical Center. A graduate of the University of Phoenix and the University of Washington, he received his B.S. in business management and his master's of public administration, respectively. Mr. Tadesse has been involved in various organizations including the United Nations Association, the Seattle Pacific Science Center, and the World Affairs Council.

Anna Taylor is the Head of Hunger Reduction at Save the Children UK (SCUK) leading a team of 13 food security and nutrition specialists. She has been working for Save the Children for the past 10 years. Previously, she served as a Nutrition Adviser at SCUK providing technical support to overseas programs and a lead on policy development in nutrition within the organization. She has worked, among other countries, in Bangladesh, Uganda, North Korea, and Tanzania for Save the Children and UNICEF. Her work has focused on information systems, infant and young child feeding, social protection, emergency feeding, European donor policy, and spans humanitarian and development fields.

Graciela Teruel Belismelis, Ph.D., is a full-time professor at the Department of Economics of Iberoamericana University. She is Codirector of the Mexican Family Life Survey. Dr. Teruel received her Ph.D. in economics from the University of California, Los Angeles, in 1998 and her B.A. from Mexico Autonomous Technology Institute. She is an academic member of the National Committee for the Evaluation of Social Programs in Mexico.

Andrew Thorne-Lyman, M.H.S., is a doctoral student in the Department of Nutrition of the Harvard School of Public Health and a freelance consultant. He worked as a nutritionist for 7 years for the UN World Food Programme (WFP) where he undertook nutritional surveys and assessments, helped strengthen systems to measure the nutritional outcomes of the WFP's programs, and worked on issues related to HIV/AIDS and nutrition. Prior to this, he worked for Helen Keller International Bangladesh on the Nutritional Surveillance Project from 1997 to 2000. He has a master's degree in international health from the Johns Hopkins School of Public Health.

Hans Timmer, a Dutch national, is Director of the World Bank's Development Prospects Group. Under his management, the Prospects Group produces the World Bank's annual publications, *Global Economic Prospects*, *Global Development Finance*, and *Global Monitoring Report*, in addition to a wide range of monitoring and forecasting publications. The Prospects Group is responsible for the global macroeconomic forecasts of the World Bank and focuses on cross-border flows to developing countries, from trade and financial flows to remittances and migration. With long-term scenario analysis, such structural issues as trade agreements, climate change policies, migration, global income distribution, and policies aimed at meeting the Millennium Development Goals are being analyzed. Before joining the Bank in May 2000, he was head of international economic analysis at Central Planning Bureau (CPB) for 10 years. In this role, he supervised the development of two world models: a long-term model of the world economy, and an econometric medium-term model of Organisation for Economic Co-operation and Development (OECD) economies. He has had vast experience working with the European Commission, Intergovernmental Panel on Climate Change, and the OECD, as well as with the Indian Planning Commission and the Chinese Academy of Social Sciences. He has participated in international modeling groups like LINK and GTAP. Mr. Timmer studied econometrics at Erasmus University, Rotterdam. He has been a researcher at the University of Lodz in Poland and at the Netherlands Economic Institute.

Ricardo Uauy, M.D., Ph.D. *(Planning Committee Member)*, is a Professor of Public Health Nutrition at the London School of Hygiene and Tropical Medicine and a Professor of Nutrition and Pediatrics at the University of Chile. He joined the Institute of Nutrition and Food Technology of the University of Chile (INTA) in 1977 as an Associate Professor, and in 1981 he became Professor of nutrition and pediatrics. He has directed INTA's training programs, the Clinical Research Center, the Division of Human Nutrition and Medical Sciences, and was Resident Coordinator for UN University activities at INTA. He headed the area of human nutrition and medical sciences and directed the Clinical Nutrition Unit; in 1994 he became INTA's Director, and was re-appointed in 1998. From 1985 through 1990 he was Associate Professor of Nutrition and Pediatrics at the Center for Human Nutrition, University of Texas Southwestern Medical Center at Dallas. Dr. Uauy is board-certified (USA) in pediatrics and in neonatal-perinatal medicine. He has served as President of the Chilean Nutrition Society and has participated as an expert to WHO/FAO. He was a member of the NIH Nutrition Study Section and is a member of the Scientific Advisory of the Novartis Foundation. He has contributed more than 250 scientific publications on various aspects of human nutritional needs in health and disease with an emphasis on neonatal nutrition, and he has coedited three books. He is on the editorial boards of *Early Human Development*, *Nutritional Biochemistry*, *Journal of Pediatrics*, and the *Journal of Pediatric Gastroenterology and Nutrition*. He was a member of the UN ACC/SCN Advisory Group in Nutrition (AGN) and chairman of the AGN for 1997–2000. He was elected as a member of the International Union of Nutritional Sciences (IUNS) council in 1997 and chosen as president-elect in 2001. He received the McCollum award presented by the American Society for Nutritional Sciences in 2000 and was inducted as member of the Chilean Academy of Medicine in 2002. He received his M.D. degree from the University of Chile in 1972 and his Ph.D. in nutritional biochemistry from the Massachusetts Institute of Technology in 1977.

Keith P. West, Jr., Dr.P.H., M.P.H., R.D. *(Planning Committee Member)*, is the George G. Graham Professor of Infant and Child Nutrition and the Director of the Center and Program for Human Nutrition in the Department of International Health of the Johns Hopkins Bloomberg School of Public Health. Dr. West has worked in international public health nutrition for over three decades, concentrating on the epidemiology and prevention of vitamin A and other micronutrient deficiencies and their health consequences in children and women of reproductive age in Southern Asia (including Bangladesh, Nepal, Indonesia, Thailand, and the Philippines), the Western Pacific (Micronesia and Marshall Islands) and Africa

(Malawi and Zambia). Before entering academia, Dr. West was trained as a Registered Dietitian in the U.S. Army Medical Specialist Corps, attaining the rank of Major, and served from 1971 to 1976 as a clinical dietitian at Walter Reed Army Medical Center in Washington, DC, U.S. Army Hospitals at Ft. Dix, NJ, and on Okinawa, Japan, and as a nutrition consultant in the Office of the Surgeon General in the Pentagon from 1980 to 1984. He served as a field nutritionist in Bangladesh with Concern (Ireland) Worldwide from 1976 to 1979 where he set up child feeding programs. Dr. West has been on the faculties of the School of Medicine and Public Health at Johns Hopkins since 1982, during which time he established population nutrition research sites in Aceh, Indonesia, the terai of Southern Nepal, and in Northern Bangladesh, and has overseen the conduct of micronutrient intervention trials among nearly ~100,000 pregnant women and a similar number of infants and preschool aged children to assess impacts on survival, morbidity, growth, development and other health outcomes. He has published more than 160 research papers and scientific reviews, including a book entitled Vitamin A Deficiency: Health, Survival and Vision with Alfred Sommer in 1996. He served on the Steering Committee of the International Vitamin A Consultative Group (IVACG) from 1994 to 2006 and current sites on the Micronutrient Forum Steering Committee. Dr. West was the recipient of the International Nutrition Award from the American Society of Nutrition in 2007.

Derek Yach, M.B., Ch.B., M.P.H., is the Senior Vice President of Global Health Policy at PepsiCo where he leads the Global Human Sustainability Task Force and engagement with major international policy, research, and scientific groups. Previously, he has headed global health at the Rockefeller Foundation, been Professor of Public Health and head of the Division of Global Health at Yale University, and is a former Executive Director of the World Health Organization (WHO). Dr. Yach has spearheaded efforts to improve global health. At WHO he served as Cabinet Director under Director-General Gro Harlem Brundtland. Dr. Yach helped place tobacco control, nutrition, and chronic diseases such as diabetes and heart disease prominently on the agenda of governments, nongovernmental organizations, and the private sector. He led development of WHO's first treaty, the Framework Convention on Tobacco Control, and the development of Global Strategy on Diet and Physical Activity. Dr. Yach is a South African national. He established the Centre for Epidemiological Research at the South African Medical Research Council, which focused on quantifying inequalities and the impact of urbanization on health. He has authored or coauthored more than 200 articles covering the breadth of global health issues. Dr. Yach serves on several advisory boards including those of the Clinton Global Initiative, the World Economic Forum, the Oxford Health Alliance, and Vitality USA. Dr. Yach received his medical degree from the University of Cape Town Medical School, and his master's degree in public health from the Johns Hopkins University School of

Hygiene and Public Health. Further, he has an honorary doctorate in science from Georgetown University.

Michael E. Zeilinger, M.D., M.P.H., serves at the U.S. Agency for International Development (USAID) as Chief of the Nutrition Division in the Bureau for Global Health. His work at USAID includes technical and managerial oversight of food security, malnutrition, child blindness, and broad health research programs. Dr. Zeilinger is also responsible for USAID's Child Survival and Health Grants Program. Prior to his work with child health, malnutrition, and food security, he served as the manager and senior advisor to USAID's Infectious Disease Strategic Objective Team, focusing on malaria, tuberculosis, and neglected tropical diseases. Dr. Zeilinger also served as a member of the U.S. Delegation Expert Team on Health and Infectious Disease for the G8 Summit in St. Petersburg, Russia, in 2006. Preceding his work at USAID, Dr. Zeilinger served as the Central Asian Regional Director for Project HOPE, working on public- and private-sector-funded tuberculosis control projects in Kazakhstan, Uzbekistan, Kyrgyzstan, and Turkmenistan. His work in central Asia also included humanitarian assistance and child survival programs. Prior to his tenure in central Asia, Dr. Zeilinger worked with Birch and Davis on the Department of Defense's Military Health Service System, and the Public Health Foundation on a Empowerment Zones/Enterprise Communities Health Benchmarks Demonstration Project funded by the U.S. Department of Health and Human Services. Dr. Zeilinger has managed several community health programs. He currently holds the positions of Adjunct Professor and Professorial Lecturer at the George Washington University School of Public Health and Health Services, Center for Global Health, and Professorial Lecturer at the American University's School of International Service. In addition to instructing courses, he guest lectures on a variety of topics ranging from emerging infectious diseases, HIV/AIDS, nutrition, integrated management of childhood illnesses, and monitoring and evaluation of international public health programs. Dr. Zeilinger also serves as a member of the board of directors of Jewish Healthcare International. Dr. Zeilinger is a doctor of podiatric medicine and holds a master's degree in public health (international health promotion) from George Washington University.

Appendix C

Workshop Registrants

Caroline Abla, U.S. Agency for International Development (USAID)
David Ahn, U.S. Department of State
Sada Aksartova, U.S. Government Accountability Office (GAO)
Kevin Anderson, Friends of the World Foundation
Jennifer Arthur, Ministry of Education
Susanna Baker, USAID
Brian Barrows, Naval Research Laboratory
Donna Barry, Partners in Health
Rashmi Basapur, Pan American Health Organization (PAHO)
Naomi Baumslag, Georgetown University Medical School
Abigail Beeson, CARE
Jandel Benjamin, United Planning Organization
Laura Birx, USAID
Richard Bissell, National Research Council
Ashleigh Black, Center for Global Health
Robert Black, Johns Hopkins Bloomberg School of Public Health
Dominique Blariaux, European Commission
Maurice A. Bloem, Church World Service
Marisa Boaz, American Dietetic Association
Kim Boortz, Kaiser Family Foundation (KFF)
John Boright, National Academy of Sciences (NAS)
Reena Borwankar, Academy for Educational Development (AED)
Christian Braneon, Georgia Tech
Henk-Jan Brinkman, UN World Food Programme (WFP)
Allyson Brown, Concern Worldwide

Ersilia Buonomo, University of Tor Vergata
Stacey Burch, Global Food and Nutrition
Kurt Burja, WFP
Joanne Burke, University of New Hampshire
Annina Burns, Institute of Medicine (IOM)
Xiaodong Cai, United Nations Children's Fund (UNICEF)
Katie Campbell, Friends of the World Food Program
Judy Canahuati, USAID
Alicia Carbaugh, KFF
AnnaSara Carnahan, LMI
Jill Ceitlin, PAHO
Kimberly Cernak, USAID
Deirdra Chester, U.S. Department of Agriculture (USDA)
Mike Chimbucimbu, Action on Tuberculosis and Health Foundation
Parul Christian, Johns Hopkins University
Eunyong Chung, USAID
Christina Clark, Save the Children
Bruce Cogill, A2Z Project, AED
Edward Cooney, Congressional Hunger Center
J.B. Cordaro, Mars, Inc.
Melissa Covelli Derry, Bill & Melinda Gates Foundation
Patricia Cuff, IOM
Karen D'Attore, Friends of the World Food Program
Patricia N. Daniels, Africa Bureau/Sustainable Development
Elizabeth Dawes, George Washington University
Aweke Teklu Debaba, Hawassa University
Jorge Deleon, Universidad de San Carlos de Guatemala
Donna Derr, Church World Service
Leanna Diedhiou, Women Thrive Worldwide
Tania Dutta, NAS
Lisa Eakman, Chicago Council on Global Affairs
Brian Egger, GAO
Fernanda Ellenberg, Robert F. Kennedy Center for Justice & Human Rights
Nancy Emenaker, National Cancer Institute
Martelle Esposito, Community Food Security Coalition
Jessica Fanzo, Millennium Villages Project
Margie Feris-Morris, FMA, LLC
Christy Forster, George Washington University
Rebecca Xiaoxiao Fu, PAHO
Ellen Girerd-Barclay, Action Against Hunger
Brittany Goettsch, Center for Strategic and International Studies (CSIS)
Abby Goldstein, USAID

Marselha Gonçalves Margerin, Robert F. Kennedy Center for Justice & Human Rights
Ruben Grajeda, PAHO
Fred Grant, Land O'Lakes
Kajal Gulati, International Food Policy Research Institute (IFPRI)
Prea Gulati, George Washington University
Antoinette Habinshuti, Women Thrive Worldwide
Jean-Pierre Halkin, European Commission
Paige Harrigan, Save the Children USA
Ellen Harris, Beltsville Human Nutrition Research Center
Gail Harrison, UCLA School of Public Health
Syed Saqib Hassan Rizvi, Pakistan Women Welfare Forum, Rawalpindi, Pakistan
Richard Hatzfeld, APCO Worldwide
Shannon Hayden, CSIS
Khosrow Heidari, South Carolina Department of Health and Environmental Control
Mary Hennigan, Catholic Relief Services
Amanda Hinkle, International Fund for Agricultural Development
Kelly Horton, Connect Nutrition
Minha Husaini, Leuser International Foundation
Tarek Hussain, UNICEF Egypt
Laura Iiyama, journalist
Paul Isenman
Cheryl Jackson, USAID
Yvonne Jackson, Administration on Aging
Ebony James, USDA/FNS Child Nutrition Division
Louise Jordan, IOM
Melissa Joy, USAID
Allan Jury, WFP
Samuel Kahn, Nutrition and Food Consultant
Caroline Kanaiza, Medair in South Sudan
Dominique Karas, Johns Hopkins Bloomberg School of Public Health
Maria Kasparian, Edesia, LLC
Randee Kastner, Center for Medical Technology Policy
Jennifer Kates, KFF
Bridget Kelly, IOM
Peter Kingori, WFP, Somalia
Amanda Klasing, Robert F. Kennedy Center for Justice & Human Rights
Rebecca Klein, Johns Hopkins Center for a Livable Future
Rolf Klemm, Johns Hopkins University
Brigette Knight, Beltsville Human Nutrition Research Center
Stacey Knobler, Fogarty International Center/National Institutes of Health

Vivica Kraak, Save the Children
Brian Kriz, Save the Children US
Kathleen Kurz, AED
KD Ladd, International Medical Corps
Karin Lapping, Save the Children, US; Alive and Thrive
Carell Laurent, USAID
Robert Lawrence, Johns Hopkins Center for a Livable Future
Yasemin Lawson, KFF
Richard Leach, Friends of the World Food Program
Karen LeBan, CORE Group
Kaia Lenhart, GMMB
Ari Levitus, PATH
Denise Lionetti, PATH
Donas Lwanga
Thulani Maphosa, Swaziland National Nutrition Council
Lex Matteini, Runyon Saltzman & Einhorn
Molly McKee, St. Mary's College of Maryland
Phillip McKinney, Food and Agriculture Organization of the UN
Carla Mejia, Institute of Food Technologists
Dasha Migunov, John Snow International
Evelyn Minor, Downtown Clusters Aging Services
Regina Moench-Pfanmner, GAIN
Lillie Monroe-Lord, University of the District of Columbia
Mark Moore, The Kibo Group
Cecilia Morales, Barros International, Ltd.
Peter Morris, USAID Office of U.S. Foreign Disaster Assistance
Rebecca Morrison, WFP
Don Morton, U.S. Department of State
Eric Muñez
Melissa Musiker, Grocery Manufacturers Association
Sorrel Namaste, National Institutes of Health
Roni Neff, Johns Hopkins Center for a Livable Future
Terra Newman, USAID
Rachel Nugent, Center for Global Development
Immaculate Nyaugo, Ministry of Health
Adebayo Ogunlade, North-West University
Toluope Olofinbiyi, IFPRI
Eleese Onami, Providence Hospital
Maria Oria, IOM
Ulyana Panchishin, GAO
Sohyun Park, Johns Hopkins School of Public Health
Angelique Paulussen, Royal DSM
Danielle Peregoy, American Society for Nutrition

Katrine Pritchard, GMMB
Zachary Pusch, Georgetown University
Donya Rahimi, Chemonics
Jennifer Rainey, George Washington University
Becky Ramsing, REACH Global
Rahul Rawat, IFPRI
Jennifer Rigg, Save the Children
Bob Roehr, British Medical Journal
Kendra Rowe, Capital Area Food Bank
Sarah Sandison, USAID
Salvador Sarmiento, Robert F. Kennedy Center for Justice & Human Rights
Morton Satin, Salt Institute
Sara Schaefer, SUSTAIN
Katey Schein, USAID
Nina Schlossman, Global Food and Nutrition
Divya Selvakumar, Baltimore Community City College
Magdalena Serpa, AED
Eleonore Seumo, AED
Nida Shakir, Senator Richard Durbin
Anne Spica, Chemonics
Desiree Stapley, USDA Food and Nutrition Information Center
Jessica Steele, PAHO
Heather Stone, Global Action Against Poverty
Hope Sukjan
Shelly Sundberg, Bill & Melinda Gates Foundation
Stephan Tanda, Royal DSM
Charles Teller, Population Reference Bureau
Phillip Thomas, GAO
Roshan Thomas, UNC Chapel Hill
Ange Tingbo, Africare
Cheryl Toner, CDT Consulting, LLC
Elizabeth Turner, SUSTAIN
Hassani Turner, PepsiCo, Inc.
Laura Turner, WFP
Mishel Unar, National Institute of Public Health
Laurian Unnevehr, USDA Economic Research Service
Allison Valentine, KFF
Lane Vanderslice, World Hunger Education Service
Debbie Vargas-Collins, National Institutes of Health
Meg Voorhis, Young Health Professional Society
Emily Wainwright, USAID
Derrzell Watson
Jennifer Weber, American Dietetic Association

Kristin Wedding, CSIS
Adam Wexler, KFF
Fokko Wientjes, DSM NV
Richard Williams, Church World Service
Karen Wong, American Medical Student Association
Calita Woods, WFP
Margaret Wu, APCO Worldwide
Wenying Wu, Smithsonian Institute
Allison Yates, USDA
Tenja Young, Chemonics International